Safety Principles

Chief Editor Zhao Huanjuan

Associate Editor Wang Shu Jin Longzhe

Huang Zhian Gao Yukun

Beijing

Metallurgical Industry Press

2023

Metallurgical Industry Press
39 Songzhuyuan North Alley, Dongcheng District, Beijing 100009, China

图书在版编目(CIP)数据

安全学原理=Safety Principles/赵焕娟等编著.—北京:冶金工业出版社,2023.3

ISBN 978-7-5024-9406-3

Ⅰ.①安…　Ⅱ.①赵…　Ⅲ.①安全科学—高等学校—教材　Ⅳ.①X9

中国国家版本馆 CIP 数据核字(2023)第 034002 号

Safety Principles

出版发行	冶金工业出版社	**电　　话**	(010)64027926
地　　址	北京市东城区嵩祝院北巷 39 号	**邮　　编**	100009
网　　址	www.mip1953.com	**电子信箱**	service@mip1953.com

责任编辑　张佳丽　美术编辑　吕欣童　版式设计　郑小利
责任校对　葛新霞　责任印制　禹　蕊
北京印刷集团有限责任公司印刷
2023 年 3 月第 1 版,2023 年 3 月第 1 次印刷
787mm×1092mm　1/16;13.25 印张;320 千字;199 页
定价 48.00 元

投稿电话　(010)64027932　投稿信箱　tougao@cnmip.com.cn
营销中心电话　(010)64044283
冶金工业出版社天猫旗舰店　yjgycbs.tmall.com
(本书如有印装质量问题,本社营销中心负责退换)

Preface

Under the macro strategic background of vigorously building the "The Belt and Road Initiative", the prevention and control of production safety risks plays an important role. Based on this, department of Safety and Science Engineering of the University of Science and Technology Beijing initiated the efforts to build an international English curriculum for the prevention and control of production safety risks, which is also aimed to improve the teaching quality of universities.

Safety is a comprehensive science, which integrates with natural science, social science, system science and other domains. Its content is extensive and complex. "Safety Principles" is a supporting textbook in the transformation process of "Emerging Engineering Education". It serves the needs of "Curriculum Ideological and Political Education" and attempts to consolidete localized cases such as safety theory, accident cause theory, fire and explosion prevention safety. So it functions as a teaching material to raise engineering ethics consciousness. By studying this book, students can understand the development of safety engineering industry both in China and abroad, the basic knowledge and key points, and the latest research results of the industry. This could play a critical role in the construction of safety engineering discipline.

"Safety Principles" enables students to study theoretical knowledge systematically, broaden horizons, and understand the latest international achievements. Students will obtain a solid theoretical foundation and broad professional knowledge. The ability of practice, the consciousness of innovation and the international vision can also be their strength. Besides, "Safety Principles" further responds to the demand of countries along the "The Belt and Road Initiative" for talents who are proficient in safety.

"Safety Principles" can be used as a basic textbook for foreign students in China who major in Safety Science and Engineering. It can also be used as an English course named "Safety Principles" for Chinese students. Beyond that, it can also pave the way for the internationalization of Safety Science and Engineering.

"Safety Principles" was published after more than two years of writing and many revisions. I am particularly grateful to Professor Jin Longzhe, Associate Professor Wang Shu, Professor Huang Zhian and Senior Engineer Gao Yukun for their compilation and revisions of this book. I should also express my thanks to Professor Qian Xinming, Professor He Xueqiu, Professor Zhang Yinghua, Professor Huang Guozhong, associate Professor Ma Xin and associate Professor Gao Na for their revision of the book.

I am willing to give my sincere thank to the official teachers, teacher Gong Li and teachers of the graduate school for their assistance during the publishing period.

I would also like to appreciate thank the work of my graduate students Liu Jing, Liu Keqing, Wang Lexi, Xi Weidong and Zhang Haiying for their efforts in the preparation.

Thanks for the knowledge provided by the cited books. Thank you for your support for this textbook. If there are any mistakes, please do not hesitate to contact us.

Thank my lovely young gentleman Gou Xingshu and my husbond Gou Mingjiang for supporting my career and my love of work.

Thank you to the graduate textbook construction project of University of Science and Technology Beijing for funding this book.

Zhao Huanjuan

October, 2022

List of Main Symbols

A The measure of area

A Work heat equivalent

B Oxygen balance rate

Bi Biwo number

c Heat capacity, specific heat, speed of sound, speed of vacuum light

C_i Concentration

c_p Constant pressure heat capacity

C_R Hot coke residue

C_S Dimming coefficient

c_V Constant volume heat capacity

d Saturated moisture content, diameter

D Diameter

D_O Diffusion coefficient

E Activation energy

f Fuel air ratio

Fo Fourier number

G Mass flow

G_{cr} Critical mass flow

G_S Weight burning rate

h Convective heat transfer coefficient between flame and solid surface, Planck constant

H Enthalpy, convective heat transfer coefficient, thickness

I Light intensity

k Thermal conductivity

K Boltzmann constant

K_O Frequency factor

K_m Explosion index

L Distance

L_V Heat of evaporation

m Quality

M Mach number

n Molar mass of substance

Nu Nusselt number

O_I Oxygen index

p Pressure

Pr Prandtl number

q Heat transfer

q_v Latent heat of evaporation

q_x Heat flux

Q Heat effect

Q_L Low calorific value

Q_H High calorific value

Q_M Mass calorific value

Q_V Volumetric calorific value

r Activity coefficient

R Universal gas constant

R_a Lagrangian number

Re Reynolds number

R_I Explosive safety distance

S Net heat gain rate per unit solid surface, entropy

t Time

t_v Disturbance index (ignition delay)

T Temperature

T_f Membrane temperature

v Speed

V Volume

α Diffusion coefficient of solid

β Excess fuel coefficient

γ Specific heat ratio

ε Emissivity

λ Molecular thermal conductivity

μ Exothermic coefficient

ρ Density

σ Stefan Boltzmann constant

Φ Angular coefficient

φ The combustion heat of solid at ignition point is transferred to its surface

ω Reaction rate

Contents

1 Overview of Safety Science

1.1 The concept and characteristics of safety

1.1.1 The emergence of safety problems and their cognition process

1.1.1.1 The appearance safety problems

Safety problems appear in the ancient stone age, age of agricultural economy, the industrial age, etc. With the development of the times, human's cognition of safety, demand for safety, and exploration of safety science and technology are constantly progressing.

1.1.1.2 The process of human understanding of safety

Human understanding of safety experiences five stages, they are ignorance safety awareness stage, primary safety awareness stage, local safety awareness stage, system safety awareness stage and dynamic safety awareness stage.

Along with the growth, we know how to protect ourselves. When we were babies, we didn't know how to protect our own safety. When we grew up as children, we began to know how to protect ourselves. As we learn more and more knowledge, we know how to protect others.

1.1.1.3 Safety problems in modern society

Safety problems have environmental safety issues, nuclear safety issues, aviation and space accidents, traffic accident, mine and industrial disasters, etc. Each problem involves many safety principles to explain, understand, study, and advance.

For example:

(1) An explosion at a nuclear power plant in southern France on the morning of September 12, 2011 is known to have killed one person and injured four others, including one in a critical condition.

(2) *Le Figaro newspaper* reported that the explosion was caused by a fire in a storage room for radioactive waste, and that no nuclear leakage was detected after the explosion.

1.1.2 Concepts and attributes of safety

1.1.2.1 Definition of safety

Safety generally refers to the state of protection from injury, illness or death of persons; or from damage or loss of equipment, property.

Safety means health and without mishap. The Sanskrit is Sarva, meaning free from harm or intact. Merriam-Webster Dictionary defines safety as no harm, injury, or danger, no threat of harm, or no threat of harm, injury, or loss.

Definition of safety: a state of being (including a state of health) and the conditions of protection in which the body and mind are free from external (adverse) influences.

This definition includes two aspects: one is the safe state of human physical and mental existence, and the other is the objective guarantee conditions of objects, and this guarantee condition is not limited to the production process.

1.1.2.2 Attributes of safety

Maslow's hierarchy of needs have five levels of demand from low to high: physiological needs, social needs, needs for esteem, knowledge needs, aesthetic needs and self-actualization needs, shown in Figure 1-1. As the most basic need of human being, safety also has the natural attribute and social attribute accordingly.

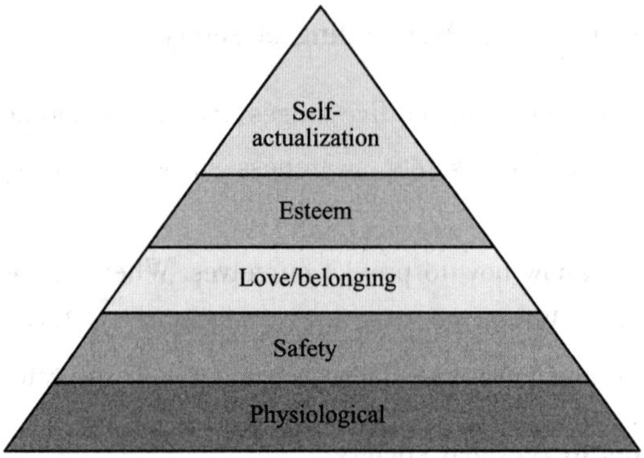

Figure 1-1 Maslow's hierarchy of needs

(1) Safety is the physical and psychological needs of people, or the desire for life and birth determines the sense of self-protection, which is the active factor of safety existence.

(2) Mankind's helpless to natural disaster is as well as metabolism and the law of life and death. It lets mankind to must put life safety on the agenda. Although this is a passive factor, it is combined with the former active factor, which can decide that safety is the eternal theme of human life, survival and progress since ancient times.

1.1.3 Basic features of safety

The necessity and universality of safety: randomness of safety, relativity of safety, safe local stability, economy of safety, safety complexity and safe sociality. Objectively, the state of "absolute safety" does not exist. There is no such thing as absolute safety in the world, and everything contains unsafe factors and has certain risks.

Here we give an example to analysis the necessity and universality of safety.

At 22:51 on August 12, 2015, a fire and explosion accident occurred in the dangerous goods warehouse of Ruihai Company in Tianjin Port, Binhai New Area, Tianjin. The total explosion energy in this accident was about 450 tons of TNT equivalent, resulting in 165 people killed, 8 missing, 798 injured, 304 buildings, 12 428 commercial vehicles and 7 533 containers damaged. The direct cause of the accident was the nitrocellulose in the container of the warehouse which accumulated heat and spontaneous combustion, at last caused the large-scale combustion of adjacent hazardous chemicals and led to the explosion of nearby ammonium nitrate and other hazardous chemicals. The indirect reason is that Ruihai company seriously violated the law and the local departments concerned did not perform their duties properly.

So the occurrence of casualty accidents is not an isolated event. Although the injury may occur suddenly at a certain moment, it is the result of a series of events. Potential hidden dangers events exist behind major accidents. It is consistent with Heinrich's Law: 1 : 29 : 300 (Figure 1-2). This law means: when an enterprise has 300 hidden dangers or violations, it is very likely that 29 minor injuries or malfunctions occur among them, and there will also be one serious injury or death accident.

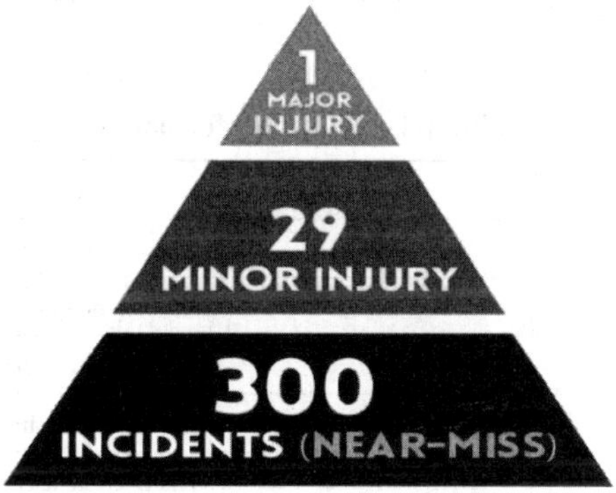

Figure 1-2 Heinrich's law

How to reduce accidents? A method is to prevent an accident happen even in the case of dangers events. In the 1950s, with the development of the world's aerospace technology, the concept of "Essential safety" was put forward. The wide acceptance of this concept is closely related to the progress of human science and technology and the understanding of safety culture.

"Essential safety": equipment, facilities or technical processes contain inherent functions that can fundamentally prevent accidents. It includes three aspects: (1) Fault-safety function; (2) Breakdown-safety function; (3) The first two safety functions should be inherent in the equipment and facilities themselves, that is, they are considered in the planning and design stage, rather than compensated after the event.

It is the process of human beings in the development process of production and life practice, from passive acceptance of accidents to active prevention in advance. The purpose of this concept is to eliminate the accident from the source and protect the safety of human beings. It shows a great progress made by human beings in safety awareness.

1.2 The emergence and development of Safety Science

1.2.1 The birth of Safety Science

Internationally, Germany established the Westphalian Mining Joint Insurance Foundation in 1863. In 1890, the Dutch Ministry of Defense supported the establishment of research on explosion prevention technology and measuring instruments, as well as laboratories for explosive identification. By the beginning of the 20th century, many western countries had established organizations and scientific research institutions related to Safety Science.

In China, the three sciences of Academician Fan Weicheng: Natural Science, Social Science and Safety Science. As we know, science is the knowledge system of human beings to know the essence and laws of things. Actually, Safety Science and Technology is a comprehensive discipline, whose discipline system is not completely the same as that of Natural Science or Social Science. The subject information of Safety Science and Technology in china is shown in Table 1-1. With no deny, Safety Science and technology not only exists interdependent relationship between each level in this discipline, but also has close relationship with other related Natural Science and Social Science.

Table 1-1 Subject information

Code	In China
08	Engineering
0837	Safety Science and Engineering
083701	Safety Science
083702	Safety Technology
083703	Safety System Engineering
083704	Safety and Emergency Management
083705	Occupation Safety and Health

With the rapid development of science and technology progress and social, in order to reduce accidents and ensure safety, health, production condition and working environment, the science and technology that related to safety is in badly need of separating from many disciplines. Then it formed into independent branch which is different from various engineering disciplines, such as: ventilation safety, electrical safety, mechanical safety, fire blast, boiler and pressure vessel safety, industrial dust, industrial gas, noise and vibration control, mining, transportation, construction, chemical industry, aviation, aerospace, agriculture, forestry, energy, textile, food and other industry safety technology.

1.2.2 The development and current situation of foreign Safety Science

The 20th century is a period of rapid development of foreign Safety Science, which can be roughly divided into three stages:

(1) The first stage: the beginning of the century to the 1950s.

In this stage, the UK, the US, Japan and other industrialized countries set up safety professional organizations, formed safety scientific research groups, and studied accident prevention technologies and methods in industrial production. Heinrich, Greenwood and other scholars have studied the Accident-Causing theory.

(2) The second stage: 1950s to mid-1970s.

After the second world war, with the development of new weapons and equipment, aerospace technology and nuclear technology, large-scale industrial production and modernization, as well as the heavy accident occurred repeatedly, the safety technology received widely attention in the various field. At the same time, the development and application of system theory, cybernetics and information theory promoted the safety system analysis and the development of technology.

During this period, the system safety analysis and safety evaluation methods have developed, such as Fault Tree Analysis (FTA), Event Tree Analysis (ETA), Failure Mode and Effects Analysis (FMEA), Hazard operability study, Fire Explosion Index Evaluation Method, Probabilistic Risk Assessment (PRA), etc. Also, put forward accident cause theory, such as the psychological motivation theory of the accident, society and environment model, domino model, the man-machine system model and so on. Safety engineering has been widely valued and gradually applied and developed in various production fields.

(3) The third stage: from the mid-1970s onwards.

As the wide application and development of system safety analysis method and safety engineering, people gradually realize the local safety flaws and research safety technology problems in all areas scattered from the multidisciplinary. Then they develop into the comprehensively research of basic safety principle and method, from the general safety engineering technology application research to the safety science theory research, discipline system, gradually establish the safety science develop accident control theory and methods, such as the essence safety, process control and behavior control and so on. Table 1-2 shows comparison of Chinese and American mine accidents.

Table 1-2 Comparison of Chinese and American mine accidents

Year	China	America	Comparison
	The number of deaths	The number of deaths	—
2009	2631	34	77 times
2010	2433	48	55 times
2011	1973	21	94 times
2012	1384	20	69 times
2013	1067	20	53 times
2014	931	16	58 times
2015	598	12	49 times
2016	538	8	67 times

1.2.3 The development and current situation of domestic Safety Science

China's safety science technology mainly gradually developed after the founding of new China, it can be roughly divided into the following three stages.

1.2.3.1 Initial establishment phase

From the early 1950s to the late 1970s, the state implemented labor protection as a basic policy, and safety technology was developed as a part of labor protection. During this period, in order to meet the needs of the development of China's industrial production, the state set up safety technology professional research institutes such as the Labor Protection Research Institute of the labor ministry, the Labor Health Institute of the health ministry, the Safety Technology Research Institute of the metallurgical ministry, the Fushun Coal Science Research Institute of the coal ministry, the Chongqing Coal Science Research Institute of the coal ministry and so on . It has also developed heat control technology, industrial dust control technology, poison hazard control technology, noise control technology, mine safety technology, mechanical and electrical safety technology, personal protective equipment and safety detection technology, etc.

1.2.3.2 Stage of rapid development

With the needs of reform and opening up and modernization, China's safety technology has developed rapidly, which is mainly reflected in the following aspects.

(1) Research institutes, centers and other research institutions engaged in safety science and technology research have been established. It has built more than 50 safety science and technology research institutes and centers, with more than 6000 professional scientific and technological persons. In particular, since the formal establishment of the Chinese Society for Science and Technology of Labor Protection in September 1983, the construction of the discipline system and professional education system of safety science and technology has been strengthened.

(2) A multi-level professional education system for Safety Science, Technology and Engineering has been established. In 1984, the Ministry of Education listed Safety Engineering in the Catalogue of Engineering Majors in Colleges and Universities. Chinese scholars such as Liu Qian put forward the assumption of establishing the discipline system of Safety Science and the architecture of Safety Science and Technology. In 1986, master's and doctor's degrees in Safety Technology and Engineering were set up in some universities, which formed a complete degree education system in the field of safety. According to incomplete statistics, until the late 1980s, 42 colleges and universities across the country had set up bachelor or specialist degree in Safety Engineering and Sanitary Engineering. 5 units were awarded the Safety Technology and Engineering disciplines (professional) master's degree by the academic degrees committee of the State Council. And 2 units were awarded doctor's degree. Safety Science and Technology of China education system has been preliminary formed. There are more than 70 large labor protection education centers throughout the country, and more than 2000 labor protection publicity and education departments in enterprises.

1.2.3.3 New stage of development

Since the 1990s, China's safety science and technology has entered a new period of development, mainly reflected in the following aspects.

(1) National standards discipline classification and code (GB/L 13745—92) lists Safety Science Technology as the First-class disciplines. In 2011, the Safety Science and Engineering was approved to be added as the First-class disciplines of the education ministry.

(2) The eighth Five-year Plan (1991-1995) and the ninth Five-year Plan (1996-2000) have included Safety Science and Technology projects.

(3) Safety Engineering series professional and technical personnel title evaluation list is provided.

(4) 1994, the labor ministry started to prepare *the Ninth Five-year Plan for the Development of Safety Science and Technology and the Outline of the long-term Goals for* 2010.

(5) The Occupational Health and Safety Management System (OHSMS), Health Safety and Environmental Management System (HSE) and other international advanced modern safety management methods were studied and applied.

(6) International exchanges and cooperation in safety science and technology have become more extensive. In particular, meetings such as the International Safety Production Forum successfully held in China shown China has more and more influential in the field of international safety.

(7) At the beginning of the 21st century, the development of safety science and technology research and professional education is more rapid. The listing of China Academy of Work Safety science is an important symbol, and various industries, fields and local have set up safety science research institutes (technology center). The Safety Technology and Engineering disciplines in China University of Mining and Technology and University of Science and Technology Xi'an were approved as national key disciplines in early 2002. In 2007, the Academic Committee of the State Council conducted the assessment, evaluation and application of the state key disciplines, and added the Safety Technology and Engineering of University of Science and Technology Beijing and Central South University as the state key disciplines. By 2011, China had more than 130 colleges and universities offering undergraduate programs in Safety Engineering, more than 20 colleges and universities offering doctoral education in Safety, more than 50 colleges and universities offering master's degree education in Safety, and more than 50 colleges and universities offering master's degree education in Safety Engineering.

1.2.4 Discipline construction

(1) Discipline classification and code—Putting "Safety Technology" in the specialized first-level discipline program and dividing it into five second-level disciplines.

(2) China Library Classification—Safety Science and Environmental Science, and classified as X.

(3) China Association for Science and Technology-China Occupational Safety and Health

Association, China Fire Protection Association, China Insurance Society.

(4) Undergraduate professional directory and postgraduate degree.

1.2.4.1 GB/T 13745—2009 Discipline classification and code in China

(1) 620 Safety Science Technology;
(2) 62010 Basic subject of Safety Science and Technology;
(3) 6201005 Safety Philosophy;
(4) 6201007 Safety History;
(5) 6201009 Safety Scientology;
(6) 6201030 Disaster Science;
(7) 6201035 Safety Science;
(8) 6201099 Basic disciplines of Safety Science and Technology other disciplines;
(9) 62021 Safety Social Science;
(10) 62023 Safety Material Science;
(11) 62025 Safety Ergonomics Science;
(12) 62027 Safety System Science;
(13) 62040 Safety and Health Engineering Technology;
(14) 62060 Safety Social Engineering;
(15) 62070 Department Safety Engineering theory;
(16) 62080 Public Safety;
(17) 62099 Other disciplines of Safety Science Technology;
(18) 62030 Safety Engineering Science;
(19) 6203005 Safety Engineering theory;
(20) 6203010 Fire Science and Fire Control Engineering;
(21) 6203020 Explosion Safety Engineering;
(22) 6203030 Safety Equipment Engineering;
(23) 6203035 Safety Mechanical Engineering;
(24) 6203040 Safety Electrical Engineering;
(25) 6203060 Safety Ergonomics;
(26) 6203070 Safety System Engineering;
(27) 6203099 Other disciplines of Safety Engineering Technology Science.

1.2.4.2 Catalogue of Subjects and Specialties (Master, Doctor)

(1) 08 Engineering;
(2) 0819 Mineral Engineering;
(3) 081901 Mining Engineering;
(4) 081902 Mineral Processing Engineering;
(5) 081903 Safety Technology and Engineering;
(6) 0837 Safety Science and Engineering;

(7)0857 Sources and Environment(including Safety Engineering, Master degree).

1.2.4.3 Undergraduate Major Catalogue(2020)

(1)0829 Safety Science and Engineering;
(2)082901 Safety Engineering;
(3)082902T Emergency Technology and Management;
(4)082903T Occupational Health Engineering.

1.2.4.4 China Library Classification(2010)

(1)X Environmental science, safety science;
(2)X1 Basic theory of environmental science;
(3)X2 Society and environment;
(4)X3 Environmental protection management;
(5)X4 Disaster and its prevention;
(6)X5 Environmental pollution and its prevention;
(7)X7 Waste treatment and comprehensive utilization;
(8)X8 Environmental quality assessment and environmental monitoring;
(9)X9 Safety science.

1.3 The definition and discipline system of Safety Science

1.3.1 Definition of Safety Science

German scholar Kuhlmann has made such an elaboration on Safety Science: "The main purpose of Safety Science is to keep the harmful effect of the technology used to be absolutely minimized, or at least to limit the harmful effect within the allowed scope. The specific function of the whole science is to acquire and summarize the relevant knowledge, and to introduce the relevant knowledge of discovery and acquisition into Safety Engineering."

German scholar Kuhlmann and Belgian scholar J. Gerson defined Safety Science as follows: Safety Science studies the relationship between human, machine and environment with the goal of establishing a balanced common ecology among the three.

In 1985, the Chinese scholar Liu Qian defined safety science in *Chinese Journal of Safety Science* as: Safety Science is a specialized research which people's physical and mental safety in production and other activities (including safety, health, comfortable, happy and enjoy), in order to protect activists and activity ability and safeguard the efficiency of the its activities across categories, comprehensive transection science.

"Safety Science" is a safety knowledge system generated from the perspective of human being's protection from external dangerous and harmful factors, starting from the creation of human safety conditions in the process of production, life and survival, and based on the summary of the whole objective world and its laws. The research of "Safety Science" aims to ensure people's personal safety and health in production and life, protect equipment and property from harm, reveal the objective laws of safety, and provide safety discipline theory, application theory and professional theory. The organizational structure of the Safety Science and Engineering Assessment Group of the State Council is shown in Table 1-3.

Table 1-3 Safety Science and Engineering Assessment Group of the State Council

Identity	Name	College
Convener	Fan Weicheng	Tsinghua university
Members	Li Shugang	University of Science and Technology Xi'an
	Zhang Laibin	China University of Petroleum
	Song Shouxin	Beijing Jiaotong University
	Jin Longzhe	University of Science and Technology Beijing
Secretary general	Shen Shifei	Tsinghua university

1.3.2 The research content and object of Safety Science

1.3.2.1 The research content of Safety Science

A The philosophical basis of Safety Science

Marxist philosophy is the world outlook and methodology of human to know the world and solve problems, establish the philosophy of safety science is the basis of the research on safety. Only to establish the correct safety concept and methodology, human can correctly analyze safety issues, solve the problem of safety, establish safety law, and provide the guidance method of science in order to solve the safety problems faced by human social.

B Basic theory of Safety Science

Human beings are faced with various safety problems, everyone has its own special rules, but in the nature of safety, there are common rules. The basic theory of Safety Science is to establish the common law of safety essence under the guidance of Marxist philosophy by applying the achievements of various basic disciplines at the present stage (such as dynamic theory, accident cause theory, safety man-machine engineering, safety rheology-mutation theory, etc.).

The research field: the production field (Production safety, Labor protection), the life field (Traffic safety, Fire and family safety), the survival field (Control and treatment of industrial pollution, Disaster control and prevention).

C Basic theory of Safety Science and Technology

(1) Dynamic theory is a theory to determine the position and direction of labor safety and health work in social production, and to guide and promote the regular development of labor safety and health work.

(2) Accident-Causing Theory studies the causes and mechanism of industrial accidents and occupational hazards, and seeks for the law under which industrial accidents and occupational hazards willoccur.

(3) Ergonomics theory studies how to maintain a coordinated, safe, comfortable and efficient human-machine relationship between human and the operating environment and machines and equipment.

(4) From the perspective of "everything flows and everything changes", the rheology-mutation theory reveals the movement process of the contradiction between safety and danger in the development process of things.

Every kind or every new machine put into use breeds new failures or accidents. The process from safety to danger also has the basic characteristics as shown in the Figure 1-3.

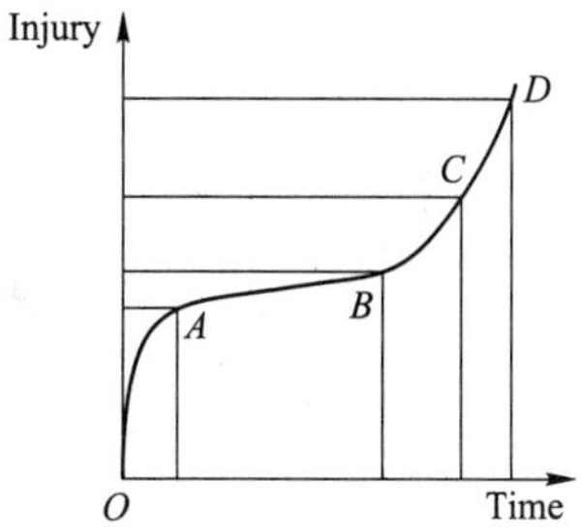

Figure 1-3 Safety rheology-mutation process

The *OA* section is the part running and wearing section of the machine at the initial stage of putting the machine into use, which is characterized by deceleration and increased wear.

AB section refers to the constant-speed wear and aging section of the machine after initial wear. In this stage, the failure rate of the machine is low and the operation is stable.

BC segment is the accelerated wear and aging segment related to component life. In this stage, the machine failure rate increases and the wear and aging amount increases dramatically.

Point *C* refers to the critical wear and aging amount of the machine from safety rheology to safety mutation transformation. This wear and wear curve has also been confirmed by tests.

D The applied theory and technology of Safety Science

Studying the application theory and technical problems of Safety Science, including safety system engineering, safety control engineering, safety management engineering, safety information engineering and safety theory and technical problems in various professional fields.

E The economic law of Safety Science

The basic theory of safety economy, law of economic loss of occupational injury accidents, safety benefit evaluation theory, safety technology economic management and decision-making theory, etc.

F The establishment of the theoretical model of production safety stage development

Trends in occupational safety and health deaths in the UK of 1917-2004 as Figure 1-4. Figure 1-5 demonstrates the changing trend of Japanese production safety and economic and social development stage. The changing trends of coal mine deaths in the U. S. from 1880 to 2003 is displayed in Figure 1-6. From the development trend curves of the above countries, different stages of economic and social development correspond to different stages of production safety development. The relationship between production safety and the changes of economic and social development stages as Figure 1-7.

1.3.2.2 The research object of Safety Science

The safety system of Safety Science research is a composite system composed of Man (M), Technology (T) and Environment (E). According to system theory, security system can be regarded

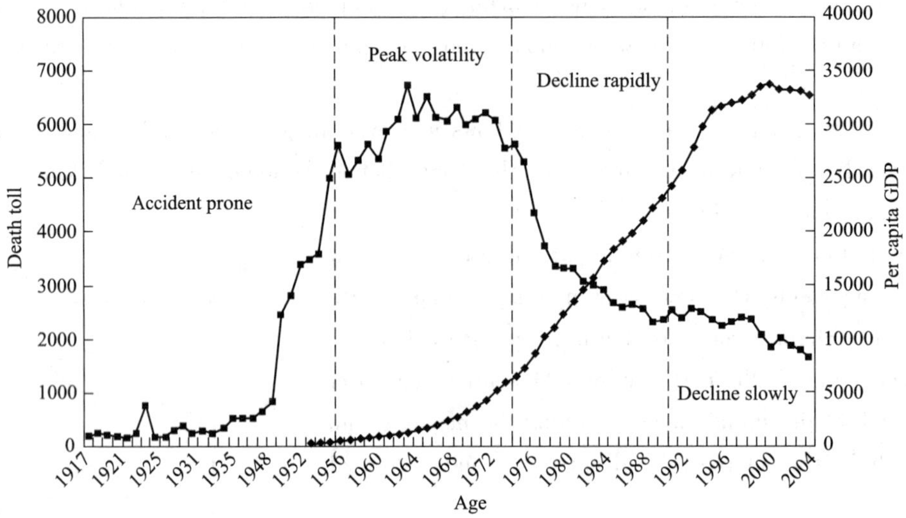

Figure 1-4 Trends in occupational safety and health deaths in the UK, 1917-2004

—■—Death toll; —◆—Per capita GDP

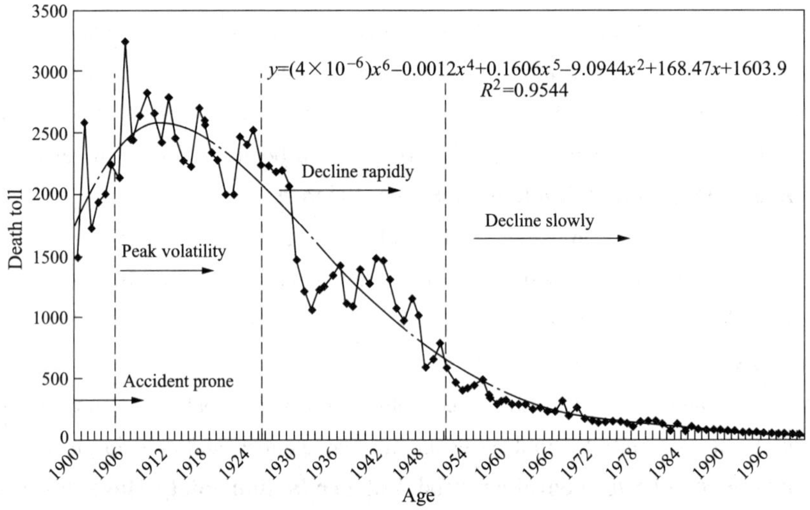

Figure 1-5 The changing trend of Japanese production safety and economic and social development stage

—◆—Coal mining deaths in the U. S. ; - · - Trends in coal mining deaths in the U. S.

as a complex system, namely MET system. This system outlines the research object of Safety Science which is shown in Figure 1-8. The MET system summarizes the research object of Safety Science, that is, the safety system consists of 7 basic subsystems, each of which puts forward the following security propositions:

(1) M: Safety psychology, safety physiology, safety education, safety behavior;

(2) E: Physical and chemical environment (labor and health environment, dust control, poison control, noise and vibration control, radiation protection, three wastes treatment), physical and chemical environment (social environment, social ethics, social economy, system and management);

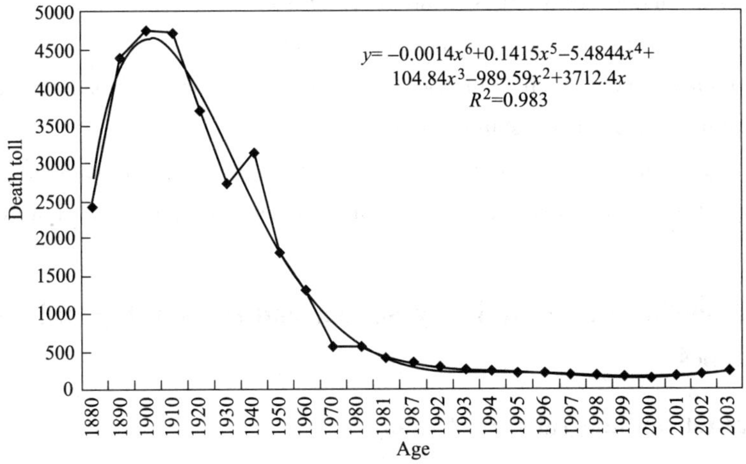

Figure 1-6 The changing trends of coal mine deaths in the U. S. , 1880-2003

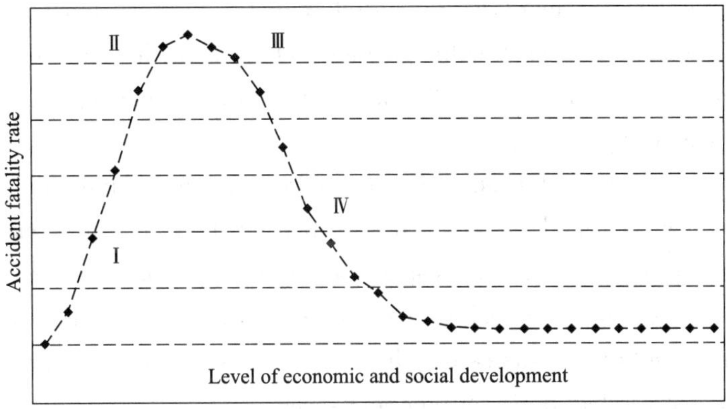

Figure 1-7 The relationship between production safety and the changes of economic and social development stages

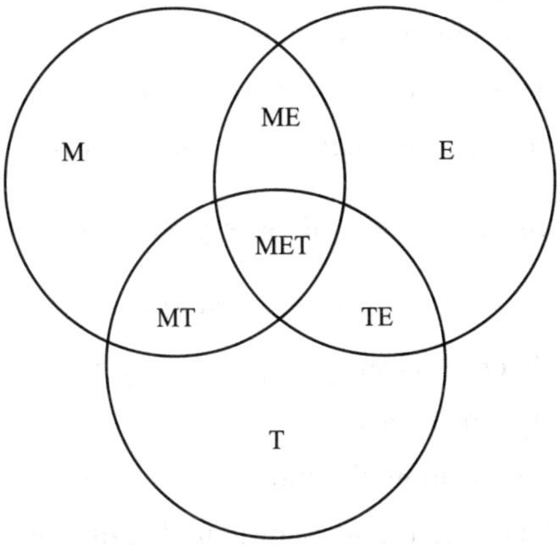

Figure 1-8 Relationship

(3) T: Reliability theory (intrinsically safe), safety technology (fire prevention, explosion protection, electro-mechanic safety, transportation safety, etc.);

(4) MT: Human-mechanism system, human-machine design;

(5) ME: Relationship between human and environment, occupational pathology, environmental standards (Operating environmental standards);

(6) TE: Environmental detection, automatic alarm and monitoring, technical risk;

MET: Safety System Engineering, Safety Management Engineering, Safety Law, Safety Economics.

1.3.3 The discipline system of Safety Science and relationship with other disciplines

1.3.3.1 Constitute an integral part of safety

(1) Human, safety human body, is the theme and core of safety, is the starting point and end-result of studying all safety problems. Human being is not only the object of protection, but also the guarantee condition or hazard factor. There is no safety problem without human being.

(2) Material, a safe substance may be the guarantee of safety or the cause of harm. The field in which the existence of substances capable of safeguarding or endangering human beings is extensive and complex in form.

(3) The relationship between human and material includes the relationship between man and man as well as between man and thing and the relationship between safety man and thing. In a broad sense, it is the link between people's safety or not, including the existence space and time of people and things, as well as the mutual connection between energy and information.

(4) Safety information is the resource on which safety activities depend. It is the qualitative or quantitative expression of safety things in time and space.

The types of safety information can be divided into primary safety information and secondary safety information. Primary safety information refers to the objective safety of man-machine environment in the production process; Secondary safety information includes safety regulations, regulations, policies, safety science theories, summaries, analysis reports, etc.

Three elements of safety science: human, material, human and material.

Four elements of safety science: human, material, human and material, information.

1.3.3.2 The level of disciplinary system of Safety Science

(1) Basic theory of Social Science: Safety Philosophy (Safety Concept);

(2) Basic theory of Natural Science: basic theory of Safety Science;

(3) Technology science: Safety Engineering;

(4) Applied science: Safety Application Technology.

The discipline system of Safety Science as shown in Figure 1-9. The science system model of Safety Science discipline as shown in Table 1-4.

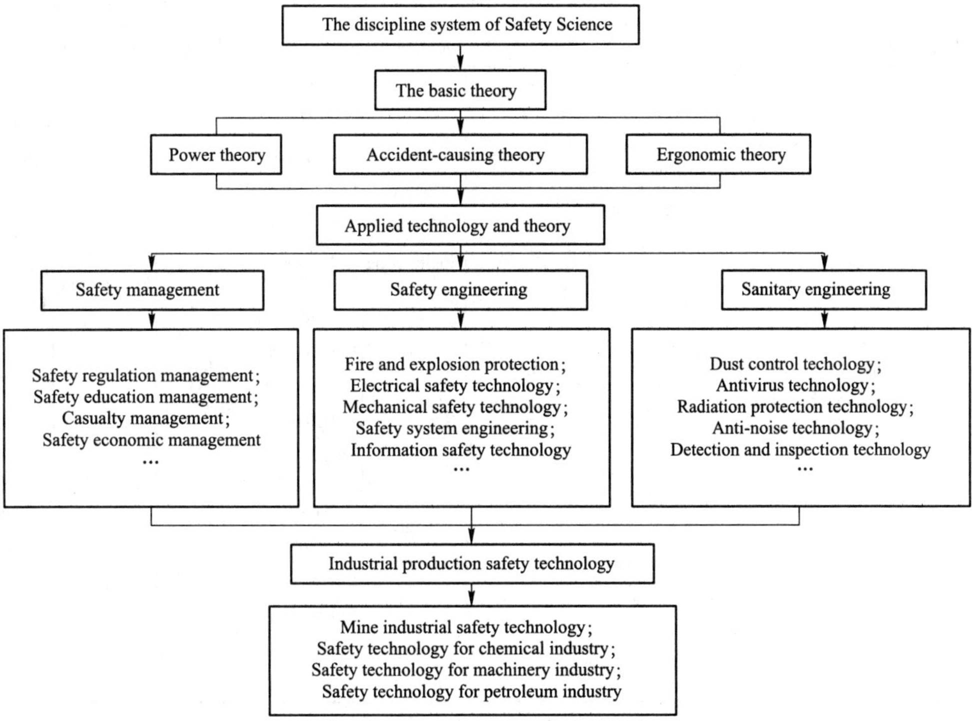

Figure 1-9 The Safety Science system diagram proposed by some experts

Table 1-4 Science system model of Safety Science discipline

Philosophy		Basic Science		Engineering Theory			Engineering and Technology		
Marxist Philosophy	Safety Concept	Safety Science	Safety Material (Material Science)	Safety Ergonomics	Safety Equipment Engineering	Safety Equipment Mechanical Engineering	Safety Engineering	Safety Equipment Engineering	Safety Equipment Mechanical Engineering
						Safety Equipment Hygiene Engineering			Safety Equipment Sanitation Engineering
			Sociology of Safety (Social Sciences)		Safety Social Engineering	Safety Management Engineering		Safety Social Engineering	Safety Management Engineering
						Safety Economic Engineering			Safety Economic Engineering
						Safety Education Engineering			Safety Education Engineering
						Safety Law			Safety Regulations
						...			...

Continued Table 1-4

Philosophy		Basic Science		Engineering Theory			Engineering and Technology		
Marxist Philosophy	Safety Concept	Safety Science	Safety Systems Science (Systems Science)	Safety Ergonomics	Safety System Engineering	Safety Operation Technology	Safety Engineering	Safety System Engineering	Full Operational Planning Technology
						Safety Information Theory			Safety Information Technology
						Safety Cybernetics			Safety Control Technology
			Safety Ergonomics (Human Science)		Safety Ergonomics	Safety Physiology		Safety Ergonomics	Safety Physiology
						Safety Psychology Engineering			Safety Psychology Engineering
						Safety Man-machine Engineering			Safety Man-machine Engineering

1.3.3.3 The disciplinary classification of Safety Science

On November 1,1992,the State Administration of Technical Supervision classified,combined and concluded into 58 first-level disciplines, 633 second-level disciplines and 2134 third-level disciplines all over 3000 disciplines in the natural and social sciences according to their principles of scientificity,practicality,simplicity,compatibility,extensibility and uniqueness.

The officially promulgated Discipline Classification and Code(GB/T 13745—92) lists "Safety Science and Technology" as a first-level discipline,consisting of 5 second-level disciplines and 27 third-level disciplines.

Among them, Safety Law (820.3080), Ventilation and Air-Conditioning Engineering (560.5520), Radiation Protection Technology (490.75) belong to the first-level disciplines of Law,Civil Engineering,and Nuclear Science and Technology.

A Subject name

(1)620. Safety Science and Technology;

(2)620.10 Basic subject of Safety Science and Technology;

(3)620.20 Safety Science;

(4)620.30 Safety Engineering;

(5)620.40 Occupational Health Engineering;

(6)620.50 Safety Management Engineering;

(7)620.99 Other disciplines of Safety Science and Technology.

B Natural Science and Social Science

(1) Science, Agronomy, Medicine, Management, Military Science, Engineering, History,

Literature, Law, Philosophy, Economics, Pedagogy.

(2) Comprehensive Science, Safety Science and Engineering Technology.

(3) Safety discipline and Engineering discipline.

(4) Safety discipline: mighty, theory, original.

Safety discipline and other disciplines support Safety Technology and Engineering (cross attribute, edge attribute). Figure 1-10 displays the comprehensive characteristics of the Satety displine.

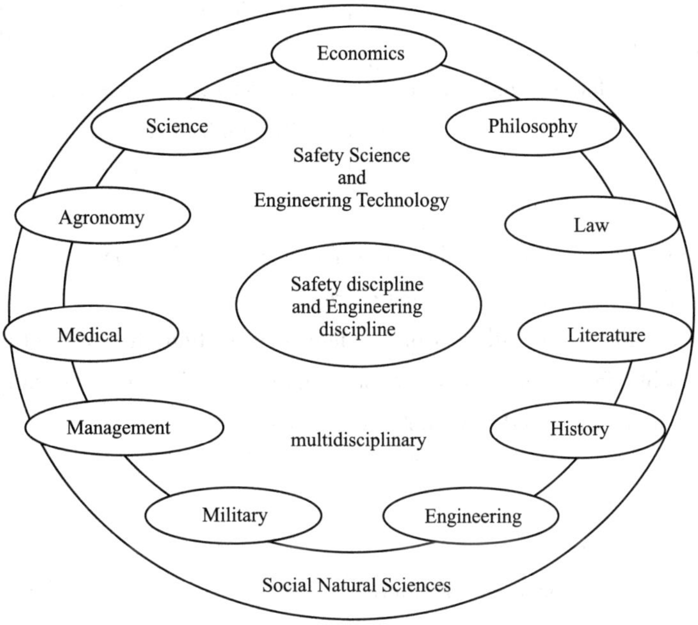

Figure 1-10 Comprehensive characteristics of the Safety discipline

The extension is fuzzy. Figure 1-11 shows the relationship between the Safety discipline and others.

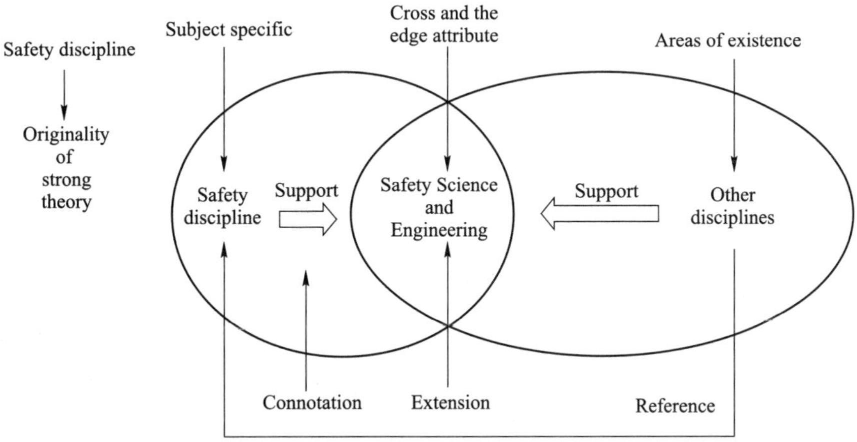

Figure 1-11 The relationship between the Safety discipline and others

1.4 The basic terminology of Safety Science

Safety
Hazard
Accident
Risk
Occupational diseases
Risk assessment
Hazard identification
Risk factors, harmful factors and potential accidents
System safety
Labor protection
Occupational safety and health
Safety information
Safety benefits
Safety objective management
Occupational health and safety management system
Performance measurement
Tolerable risk

The pollution prevention facilities in the construction project shall be designed, constructed and put into operation at the same time as the main project.

Four serious treatments are not allowed in accident investigation: the cause of the accident is not clear; those responsible for the accident are not handled; tectification measures have not been implemented; the people concerned have not been educated.

[Exercises]

1. Your definition of "Safety".
2. Why we need to study the principle of safety?

2 Basic Knowledge of Safety Science

2.1 The philosophical basis of Safety Science

A. B Maslow, an American psychologist, believes that safety is one of the basic human needs.

Maslow's hierarchy of needs have five levels of demand from low to high which are as follows.

(1) Physiological needs: food, shelter, clothing;

(2) Safety needs: physical safety;

(3) Social needs: friendships, group membership, social connection;

(4) Esteem needs: approval recognition, self-confidence;

(5) Self-actualization needs: enrich your knowledge and achieve success.

2.1.1 The unity and contradiction between safety and danger

(1) Safety and hazard are a pair of contradictory, it has all the attributes of contradiction. On the one hand, both sides oppose each other, reject each other and deny each other. The higher the degree of safety, the smaller the risk will be, and the smaller the degree of safety, the greater the danger will be. On the other hand, safety and hazard are interdependent in a unified body, and has a tendency to transform to each other.

(2) Safety is relative; danger is absolute.

The relativity of safety is reflected in three aspects:

1) The state of absolute safety does not exist, the safety of the system is relative to the danger;

2) Safety standards are equivalent to human awareness and social and economic affordability;

3) Safety is relative to human cognition.

The absoluteness of danger occurs in things at birth. In the intermediate process, the danger potential may become larger or smaller, but it will never disappear. Danger exists in any time and space of all systems. The contradictory movement, change and development of safety and danger promote the development of safety science and the enhancement of human safety consciousness.

2.1.2 The connection view and system view of Safety Science

In order to reflect the impact on safety and risk factors of the inherent regularity, Safety Science must comprehensively analyze the elements, use what we have achieved from various disciplines, analyze and synthesis the open big system, focus on grasping the main contradictions within things on the basis of comprehensive analysis, find out the objective law and the implementation way of safety.

The integrity of the system is determined by the comprehensive action of various elements, and it is the interaction and connection of various elements within the system that produce synergistic effect.

2.1.3 Qualitative change and quantitative change in safety

Quantitative change and qualitative change in philosophy are represented by rheological change and mutation in safety science. Rheology is a slow change process, mutation is the interruption in the process of rheology, is a qualitative leap.

The unity of rheology and mutation is shown in three aspects:

(1) Rheology is opposite to mutation.

If the water level of a river does not always change beyond its bed, nothing will happen. The river overflowed its bed and became a flood.

(2) The rheology and mutation are hierarchical.

If an organ is damaged in the body, it is a mutation event in a small area. For the whole human body, it is the rheology of comprehensive functions.

(3) Rheology and mutation can transform each other.

For example, the process of biological evolution is a slow rheological process, but for hundreds of years, due to human deforestation, killing of animals, use of pesticides and discharge of waste, many biological species extinction events have occurred.

2.1.4 The inevitability and contingency of safety accident

For example: the counter damper in the ventilation room of the mine ventilation system is a safety door that can be lifted up and down under normal conditions. It is inevitable to operate dexterously and freely, and it is accidental that it cannot be adjusted when disaster occurs. However, due to neglect of management, failure to maintain according to regulations, pulley rust, deformation of the frame and other reasons, such a safety measure can not play a normal role in disaster has become inevitable.

2.1.5 Safety issues of simplicity and complexity, accuracy and fuzziness

2.1.5.1 Simplicity

(1) Complex system can be decomposed into simple elements, elements and units;

(2) The relationships between the inside and outside of complex systems and the basic laws they follow are simple.

2.1.5.2 Complexity

The safety system contains infinite levels of contradictions, forming extremely complex structure and function, and has a variety of connections and interactions with the outside world.

2.1.5.3 Accuracy and fuzziness

The cognition of safety science always changes from fuzzy to precise, which is dialectical and unified. Fuzziness can explain accuracy, while appropriate fuzziness is accurate. However, the boundary of fuzzy qualitative description is too wide, which will reduce the degree of safety.

There is no precise line between safety and hazard. It is a fuzzy concept, but it can be better explained by precise Numbers.

2.1.6 Safety philosophy

(1) In terms of analyzing and understanding problems, Safety Science should:

1) Everything is based on reality.

2) To grasp the essence of things in universal relations.

3) A method to grasp the law of safety in dynamic.

4) Adopt contradiction analysis method.

(2) The essential characteristic content of Safety Science:

1) Safety Science reflects the essential safety, that is, to achieve the safety of things or systems in essence.

2) Safety Science should reflect the theoretical and scientific nature.

3) Safety Science should embody intersections and form a systematic theory.

4) Safety Science should reflect the comprehensiveness of the research object.

5) The purpose of safety science is to reflect the optimization of human, economic, environmental and technological functions.

(3) Basic features of safety: necessity and universality, randomness, relativity, local stability, economy, complexity, sociality, occult.

1) The necessity and universality of safety. Safety is a necessary prerequisite for human survival. Safety is absolutely necessary as a condition of a person's physical and mental state from which it is guaranteed. The realization of human safety is also universally needed. In all fields of human activity, we must try our best to reduce mistakes, reduce risks, and make things as safe as possible.

2) Randomness of safety. Safety depends on the coordination of people, things and the relationship between people and things, if the disorder will appear harm or damage. The existence and maintenance of safety state, location and the way of dynamic balance are random.

3) Relativity of safety. Safety standard is relative, the degree of extension of the connotation of safety and the degree of strictness of the standard depend on the range of people's physiology and psychology, the level of scientific and technological development and the political and economic status. Social ethics and safety law concept, the people's degree of material and spiritual civilization and other practical conditions.

4) Local stability of safety. It is impossible to pursue absolute safety unconditionally, especially for giant systems. However, it is possible and necessary to realize the local safety of people or the essential safety of things conditionally.

5) Economy of safety. As the first productivity, safety science and technology can not only improve production efficiency, but also play an important role in maintaining and guaranteeing the safe operation of production, human life and health. As a productivity input, it has its economic value of gift, including the safety performance of the product itself also contains the potential

economic value of safety. On the other hand, safety guarantee without danger, injury and damage is equal to the creation of economic benefits.

6) The complexity of the safety. Safety depends on the coordination of people, objects, environment and their interrelations, which actually forms the operating system of people, machine and environment, which is an open giant system combining nature and society. In safety activities, due to the leading role and essential attributes of human, including human thinking, psychological, physiological and other factors, as well as the relationship between human and society, namely human biology and sociality, the safety problem has great complexity.

7) Sociality. safety is directly related to social stability. Both man-made and natural disasters, casualty accidents in production, the transportation of the car accidents, plane crash, damage and fire, and of the family product harm to consumers, drugs and chemicals influence to human health, and even travel and entertainment small accident brings to the national economy and people's livelihood spiritual and physical harm, become the important factors that affect social stability.

8) Occult. The degree of understanding of the nature of safety and the law of movement change of all kinds of things is always limited by people's cognitive ability and technological level. Many products still have a lot of potential harm, and people still need to do in-depth research.

2.2 The theoretical basis of Safety Science

2.2.1 Safety system theory

2.2.1.1 Safety systematic engineering

Safety system engineering is a set of management procedure and method system that applies the principle and method of system engineering, analyzes, evaluates and eliminates all kinds of risks in the system, and realizes the system safety.

System engineering refers to the technology of analyzing and designing system components, organizational structure, information flow and control structure in order to better achieve system objectives.

2.2.1.2 Contents of safety system engineering

(1) System safety analysis. System safety analysis is the core of safety system engineering. The correctness of the analysis result is related to the success or failure of the whole safety work.

(2) System safety evaluation. Seek the lowest accident rate, the lowest loss and the best safety investment benefit.

(3) System safety prediction. In order to prevent and forecast possible accidents.

2.2.2 Safety information theory

Information theory is a science that applies probability theory and mathematical statistics to study the basic properties and measurement methods of information, and to study the general laws of information acquisition, transmission, storage, processing, transformation, extraction and utilization.

Its transmission and storage with a certain material as the carrier, with a certain amount of energy as the power.

According to the content of safety information, it can be divided into five categories:

(1) Production and operation information;

(2) Safety management information;

(3) Safety technical information;

(4) Occupational health information;

(5) Accident information.

2.2.3 Safety-control theory

Safety-control theory is a science that studies the regulation and control of animals, human beings and machines and even all systems.

Control refers to the adjustment of an organized system according to the changes of internal and external conditions, in order to overcome the uncertainty of the system, so that the system can maintain or reach a certain state stably.

The general steps of safety control are as follows:

(1) To establish judgment criteria and standards for production safety;

(2) Measure the deviation between the actual work safety activities and the predetermined target;

(3) Appropriate safety management, safety education and safety engineering and technology measures shall be taken to correct undesirable deviations or hidden dangers.

2.2.4 Safety psychology and behavior

Safety psychology is a science that studies the law of people's psychological activities from the perspective of protecting production safety, preventing production accidents and reducing personal injuries.

The research contents include: safety individual psychology, safety group psychology, safety management psychology, accident prevention psychology, disaster mental health, etc.

2.2.5 Safety management

Safety management starts with the inducing factors of safety problems and applies the relevant knowledge and theories of management to carry out safety production management. It uses scientific management methods and systematic and effective management mechanism to contain the occurrence of accidents, so as to achieve the ultimate goal of preventing accidents and safe production.

Safety management is a series of activities in which the manager plans, organizes, commands, coordinates and controls safety production.

Basic principles of safety management:

(1) The prevention principle is to prevent people's unsafe behavior and things' unsafe state

through effective management and technical means, so as to minimize the probability of accidents.

(2) Mandatory principle is based on the contingency of accident loss, people's "risk taking" psychology, and the irretrievability of accident loss. It adopts the means of compulsion management to control people's will and behavior, making individual activities and behaviors subject to the requirements of safety management, so as to achieve effective safety management.

2.2.6 Safety economics

Safety economics is to study the economic forms and conditions of safety, and to achieve the best safety benefits of human, technology and environment through reasonable organization, control and adjustment of human safety activities.

Research object is implemented according to the safety and economic effect of the relationship between the unity of opposites, from the theory and method research on how to make safety activities, with the best way and people's labor, life, survival and reasonable combination, finally achieve safety and economic and reasonable labor, life, survival, good comprehensive benefits to human society. Principle of minimum cost of safety input as shown in Figure 2-1. Principle of maximum benefit of safety input as shown in Figure 2-2.

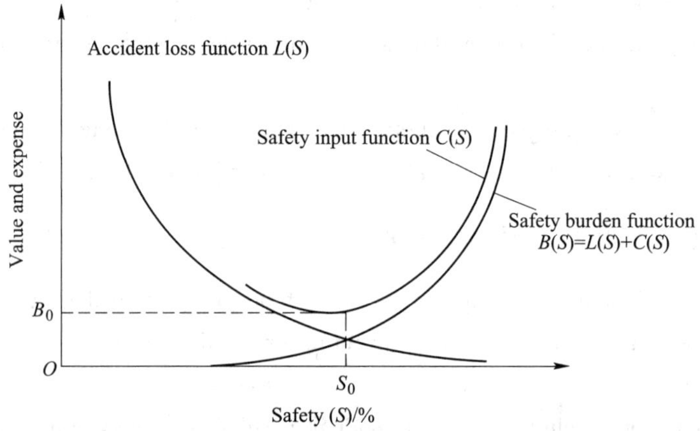

Figure 2-1 Principle of minimum cost of safety input to minimize $B(S)$

There is B_{min} at S_0, and S_0 can be obtained from the following formula: $dB(S)/dS=0$.

E_{max} Exists at point S_0, and at this point, point S_0 can be obtained by the following formula: $dE(S)/dS=0$.

When the safety is S_a and S_b, there is no safety benefit, indicating that the system safety is low or high, and the overall safety benefit is not high.

According to the "maximum benefit principle", the value of safety can be divided into three ranges:

(1) $S<S_1$, small investment, but great loss, poor comprehensive benefits, the need to improve system safety and safety benefits;

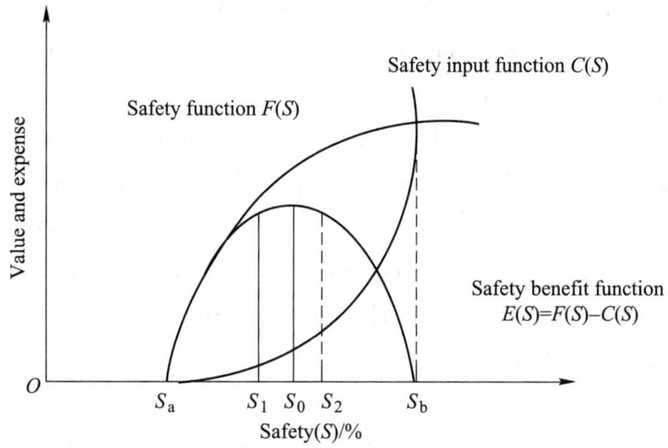

Figure 2-2 Principle of maximum benefit of safety input

(2) S is between S_1-S_2, close to S_0, which has better comprehensive safety benefits and is the preferred range.

(3) $S>S_2$, the loss is small, the safety cost is high, the comprehensive benefit is also poor, needs to make every effort to maintain the safety, enhances the safety science and technology level, reduces the cost, improves the comprehensive benefit.

2.2.7 Safety risk science

Safety risk science is a borderline subject integrating safety management, risk management and safety economics. In the science of safety risk, risk is the combination of the possibility and consequence of a specific hazardous event, and belongs to pure risk (that is, a risk with only loss opportunity). Its research contents include risk identification, risk estimation and evaluation, risk control, risk decision-making, etc.

2.2.8 Safe ergonomics

Safety and ergonomics are specializing in industrial production process of human and machine safety. That is, from the viewpoint of safety, on the basis of the theory of man-machine engineering, mechanical engineering, reliability and so on. Safe ergonomics provides safety technical information for the design and create of safety and reliable machines, and puts forward the basic safety requirements and design criteria for the machine structure design, information display and control design. In the man-machine system, human always plays a core and guiding role, while machine plays a safe and reliable role.

2.2.9 Safety culture and enterprise safety culture

Safety culture is the sum of the material wealth and spiritual wealth created by human beings to protect their physical and mental safety and health during their existence, life and production activities.

The enterprise safety culture is formed or consciously shaped in the long-term production and operation activities of the enterprise, which is accepted and followed by all the employees and has the enterprise characteristics of safety thought and consciousness, safety style and attitude, safety management mechanism and code of conduct. The enterprise's work safety goal and the enterprise's safety enterprising spirit refer to the image of enterprise safety production, such as the safe and comfortable production and living environment and conditions created to protect the physical and mental safety and health of employees, and the safety equipment and measures for disaster prevention, refuge and emergency response; The sum of various enterprise safety material factors and safety spiritual factors, such as safety values, safety aesthetics, safety psychological quality and enterprise safety style.

Basic Policy of production safety: "Safety first, prevention first". Safety management system: "State supervision, industry management, enterprise responsibility, supervision by the masses, workers abide by the rules and regulations."

Safety production regulations refer to the sum total of the laws and regulations produced in the process of production, which are used to adjust the safety and health of workers or production personnel, as well as various social relations related to the means of production, social wealth and safety protection.

2.3 The mathematical basis of Safety Science

2.3.1 Basic logical operations and logical functions

2.3.1.1 Basic logical operation

The variables of logical algebra are only 0 and 1. Represents two different logical states (for example, yes and no, true and false, high and low, something and nothing, on and off, etc.). In logical algebra, the most basic logic has three kinds: and, or, non; Use logical algebraic symbols to denote AND gate, OR gate, NOT gate. Basic logical operation is shown in Table 2-1.

Table 2-1 Basic logical operation

Name	Logic symbol	Function formula	Implication
AND Gate	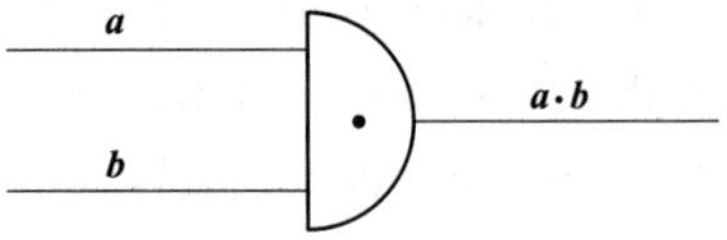	$z(a\cdot b)=a\cdot b$	$1\times1=1$ $1\times0=0$
OR Gate	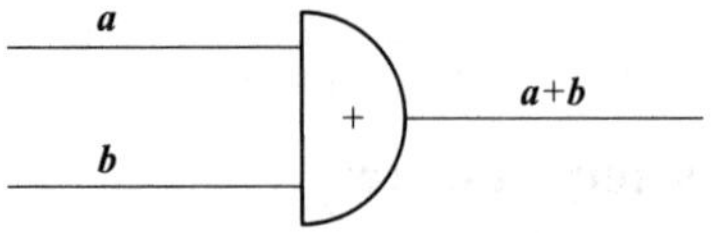	$z(a\cdot b)=a+b$	$1+1=1$ $1+0=0$ $0+0=0$
NOT Gate	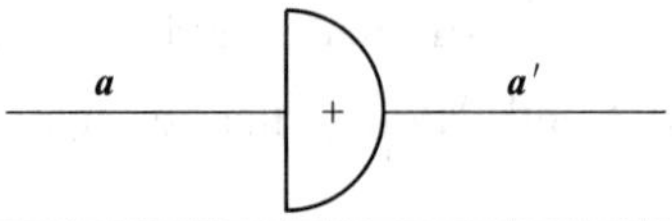	$z(a)=a'$	$a=1, a'=0$ $a=0, a'=1$

(1) And operation——abbreviation logical multiplication. Means that when the input variables are $\boldsymbol{a}$ and $\boldsymbol{b}$, output $z=\boldsymbol{a}\cdot\boldsymbol{b}$ that is, when all the conditions $\boldsymbol{a}$ and $\boldsymbol{b}$ that determine the event z are in place, the event z will occur; otherwise, it will not occur.

(2) Or operation——abbreviation logical add. Means that when the input variables are $\boldsymbol{a}$ and $\boldsymbol{b}$, output $z=\boldsymbol{a}+\boldsymbol{b}$, that is, the condition $\boldsymbol{a}$ or $\boldsymbol{b}$ that determines the event z only occurs when one or both of them have all the conditions. When neither $\boldsymbol{a}$ nor $\boldsymbol{b}$ is present, z does not occur.

(3) Not operation——abbreviation logicalnot (or logical negation). When the input variable is $\boldsymbol{a}$, output $z=\boldsymbol{a}'$ (or $\bar{\boldsymbol{a}}$), read not $\boldsymbol{a}$. That is, if the condition determining event z is $\boldsymbol{a}$, z is opposite to $\boldsymbol{a}$, and if $\boldsymbol{a}$ exists, z does not occur, and vice versa.

2.3.1.2 Logical variables and logical functions

In general, if the value of input variables $\boldsymbol{a}, \boldsymbol{b}, \boldsymbol{c}, \cdots$ are determined, then the value of the output variable z is determined. So call z is a logical function of $\boldsymbol{a}, \boldsymbol{b}, \boldsymbol{c}, \cdots$ and written as:

$$z = F(\boldsymbol{abc}\ \cdots)$$

In logical algebra, both variables and functions have only two values (0 and 1). For although there are many conditions that determine whether an event will occur, there are only two possibilities for what one condition is and what one condition is not. For events that correspond to functions, there are only two things that happen or don't happen.

(1) Logic function algorithm.

1) Idempotent rules:

$$A + A = A \quad \text{or} \quad A \cup A = A \tag{2-1}$$

2) Exchange rules:

$$A + B = B + A \quad \text{or} \quad A \cdot B = B \cdot A \tag{2-2}$$

3) combining rules:

$$A + (B + C) = (A + B) + C \quad \text{or} \quad A \cdot (B \cdot C) = (A \cdot B) \cdot C \tag{2-3}$$

4) distribution rules:

$$A + (B \cdot C) = (A + B) \cdot (A + C) \quad \text{or} \quad A \cdot (B + C) = (A \cdot B) + (A \cdot C) \tag{2-4}$$

$$(A + B) \cdot (C + D) = A \cdot C + A \cdot D + B \cdot C + B \cdot D \tag{2-5}$$

5) absorption rules:

$$A + A \cdot B = (A + A) \cdot (A + B) = A \quad \text{or} \quad B + A \cdot B = B \tag{2-6}$$

$$A \cdot (A + B) = A \cdot A + A \cdot B = A + A \cdot B = A \quad \text{or} \quad B \cdot (B + A) = B \tag{2-7}$$

(2) Event tree analysis.

A logical analysis method that uses graphics to perform deduction. It can be used to identify various possible outcomes after a hypothetical initiating event. ETA, also known as Decision Tree Method. It is a method to analyze the development and change process of "success (normal)" or "failure (failure)" of each link event from the initial cause event, and predict various possible results. That is, identification is carried out by analyzing the events of each link in the system.

(3) Fault tree analysis.

Using a logical method, according to the results of the accident that may or have occurred in the system, to find the cause, condition and law related to the accident, and identify. Fault Tree Diagram (FTD) is a kind of logical causality diagram, which uses event symbols, logic gate symbols and transition symbols to describe the causal relationship between various events in the system.

2.3.2 Random events and probability calculations

2.3.2.1 Random events

Referred to as "event". It can be regarded as a series of experiments or observations under the same set of conditions, and each experiment or observation may have more than one possible result. The exact result cannot be predicted before each experiment or observation, that is, uncertainty.

Type of event:

(1) Sub event: write for $A \subset B$.

(2) Sum event: write for $C=A+B$ (or $C=A \cup B$).

(3) Product event: write for AB (or $A \cap B$).

(4) Mutually exclusive event: write for $A \cap B=\phi$.

(5) The inverse of the event: write for $\overline{A}$ (or $\overline{B}$).

(6) Difference event: write for $C=A-B=A \cdot \overline{B}$.

2.3.2.2 Frequency

If random event A occurs m times in n trials, the ratio m/n is called the frequency of random event A, denoted as $W(A)$, expressed as follows:

$$W(A) = m \div n \tag{2-8}$$

Because of $0 \leqslant m \leqslant n$, the frequency value of random events is between 0 and 1. The frequency of inevitable events is always equal to 1, the frequency of impossible events is always equal to 0. Under a set of conditions, repeat n independent experiments, let m be the number of occurrences of event A in n experiments. If for a large number of tests, the frequency m/n oscillates around a certain value q, then q is the probability of event A occurring under this set of conditions.

Referred to these as: $Q(A)=q$, where, $0 \leqslant Q(A) \leqslant 1$; $Q(S)=1$, $Q(\phi)=0$.

2.3.2.3 Statistical definition of probability

If the probability of event A is P, it can be writtenas $P(A)=P$.

In fact, when n is sufficiently large, the frequency of event A can be used as an approximation of the probability of event A, namely:

$$P(A) = P = m \div n \tag{2-9}$$

Like the frequency, the probability of an event has the following properties: (1) $0 \leqslant P(A) \leqslant 1$; (2) $P(S)=1$; (3) $P(\varphi)=0$.

2. 3. 2. 4 Classical definition of probability

If the total number of basic events is n, and event A contains m of them, then the probability of event A is defined as:

$$P(A) = m \div n = \text{Number of the event } A \text{ included/Total number of basic events} \tag{2-10}$$

2. 3. 2. 5 Probability calculation of independent events

If the occurrence of event A does not affect the probability of event B, and vice versa, the two events are said to be independent of each other.

(1) Probability of Logical Product (Independent events are connected with gates):

$$q_{(A+B+C+\cdots+N)} = q_A \cdot q_B \cdot q_C \cdot \cdots \cdot q_N = \prod_{i=A}^{N} q_i \tag{2-11}$$

(2) Probability of logical sum (Independent events are connected by OR gate):

$$q_{(A+B+C+\cdots+N)} = 1 - (1 - q_A)(1 - q_B)(1 - q_C)\cdots(1 - q_N) = 1 - \prod_{i=A}^{N}(1 - q_i) \tag{2-12}$$

2. 3. 2. 6 Probability calculation of non-independent events

(1) The probability of a logical sum is:

$$q_{(A+B+C+\cdots+N)} = q_A + q_B + q_C + \cdots + q_N \tag{2-13}$$

(2) The probability of the logical product is:

$$q_{(A\cdot B)} = q_A \cdot (q_A \div q_B) = q_B \cdot (q_A \div q_B) \tag{2-14}$$

2. 3. 3 Reliability and probability calculation of basic events

2. 3. 3. 1 The basic concept of reliability

(1) Reliability: refers to the ability of the research object to complete the specified functions under specified conditions and within a specified time.

(2) Degree of reliability: refers to the probability that the research object will complete the specifiedfunction under the specified conditions and within the specified time. Usually denoted as R.

(3) Unreliability: refers to the probability that the research object loses the specified function under specified conditions and within a specified time. Usually marked as F.

$$R + F = 1 \quad \text{or} \quad R = 1 - F \tag{2-15}$$

(4) Failure rate: indicates the probability of failure of the research object per unit time at a certain time t.

$$\lambda(t) = \frac{1}{N_s(t)} \frac{dN_f(t)}{dt} \tag{2-16}$$

(5) Maintenance degree: It is the probability that the repairable system can complete the

maintenance under the specified conditions and within the specified time.

$$M(t) = 1 - e^{-\int_0^t \mu(t)\mathrm{d}(t)} \tag{2-17}$$

when $M(t)$ is a constant,

$$M(t) = 1 - e^{-\mu t} \tag{2-18}$$

(6) The life process of the system: MTTF (Mean Time To Failure), MTBF (Mean Time Between Failure) and MTTR (Mean Time To Repair) express the life process of the system. For repairable systems, MTBF is the average normal working time of the system, and MTTR is the average repair time. For non-repairable systems, MTTF is the average life of the system, and MTTR is the average replacement time.

(7) Effectiveness of the repairable system: when the reliability and maintenance of the system obey the exponential distribution, the effectiveness of the system is:

$$A(t) = \frac{\text{MTBF}}{\text{MTBF} + \text{MTTR}} = \frac{\mu}{\mu + \lambda} \tag{2-19}$$

2.3.3.2 Basic event probability calculation

For a general repairable system, the unit failure probability is:

$$q = \frac{\text{MTTR}}{\text{MTBF} + \text{MTTR}} \tag{2-20}$$

Because $\text{MTTR} = \frac{1}{\mu}$, $\text{MTBF} = \frac{1}{\lambda}$.

Substitute the expressions of MTTR and MTBF (2-37) and sort them out:

$$q = \frac{\lambda}{\mu + \lambda}$$

Generally speaking, MTBF ≥ MTTR, so $\lambda \ll \mu$, So there is:

$$q = \frac{\lambda}{\mu + \lambda} \approx \frac{\lambda}{\mu} = \lambda\tau \tag{2-21}$$

where $\tau = \frac{1}{\mu}$ ——time to repair.

2.3.3.3 Application examples

[Example 1] Assuming that the life x of a product obeys exponential distribution, find the average life θ.

Solution: x obeys exponential distribution, so $R(t) = e^{-xt}, t \geq 0$. Its average life:

$$\theta = \int_0^\infty e^{-xt}\mathrm{d}t = \frac{1}{\lambda}$$

$$\text{MTTR} = \int_0^\infty t\frac{\mathrm{d}\mu(t)}{\mathrm{d}t}\mathrm{d}t = \int_0^\infty t\mathrm{d}\mu(t)$$

If $\mu(t) = \mu$, then there are:

$$\text{MTTR} = \frac{1}{\mu}$$

2.3.3.4 Application examples

[Example 2] A damper needs to be repaired once every 200d, and the repair time is 8 hours each time, then:

$$MTBF = 200(d) = 200 \times 24 = 4800(h); MTTR = 8(h)$$

Throttle failure rate:

$$q = 8 \div 4808 = 0.0017$$

For an unrepairable system, the unit failure probability of the equipment is:

$$q = 1 - e^{-\lambda t} \approx \lambda t$$

Where, t is equipment running time.

[Example 3] In the above example, if it is assumed that the damper is scrapped once, the failure probability of the damper after 400 hours of normal operation is:

$$q \approx \lambda T = 1 \div 4800 \times 400 = 0.08$$

2.3.4 Several basic concepts commonly used in Safety Science

(1) Safety index: Under certain conditions, a production (or living) system, in the process of completing its functions, the acceptable level of accident losses.

For example: Million tons of steel (or coal) mortality = death toll ÷ actual output $\times 10^6$

(2) Intrinsically safe: The safety technology and safety management level of the system have reached the contemporary requirements of this department, and the system can operate more safely and reliably. Intrinsic safety does not mean that the system will never cause accidents.

Reasons:

1) The degree of intrinsic safety is relative.

2) Production is a dynamic process, and accidents are hard to predict.

3 Theory of Rheology and Mutation

3.1 Background knowledge of rheology-mutation theory

The term "Rheology" comes from ancient Greece, which means that everything flows and everything changes. "Mutation" has the meaning of radical change. It was originally proposed by Thom in the book *Structural Stability and Morphogenesis* in 1968 which means that things have undergone a qualitative and radical change. Rheology and mutation are combined to form rheology-mutation theory, which describes the entire process of things from birth to development to death.

3.1.1 Rheology-mutation theory of matter

Quality includes not only qualitative quality, but also quantitative quality. The material world is changing abruptly in constant flux.

3.1.2 The time and space view of rheology-mutation theory

The spatial extension and temporal continuity of a thing is an inherent attribute of the thing, and it must be in the same time and space as the other. And, all rheology-mutation phenomena are inseparable from the interaction of matter in space.

3.1.3 The motion view of rheology-mutation theory

The properties of things are revealed in rheology and mutation. Rheology-mutation is the process of changing one thing to another.

3.2 Basic characteristics of safety rheology-mutation

"Safety changes and mutations" is the contradiction between safety and danger in the development of things. This contradictory movement determines the safety status at all stages. Here are some typical examples to illustrate the basic characteristics of "safe rheology and mutation" (see Figure 1-3).

3.2.1 Mine disaster

3.2.1.1 Spontaneous fire

Segment *OA*: coal and oxygen start contact with oxidation.
Segment *AB*: the oxidation rate slows down and the degree of oxidation is deepened.
Segment *BD*: faster oxidation.

Point D: a spontaneous fire occurs.

Point C: the alarm point set manually.

3.2.1.2 Roof fall

Segment OA: new roadway just dug out. Deformation speed decreasing section.

Segment AB: stable section.

Segment BD: the surrounding rock pressure and other conditions change, the deformation speed increases, and roof fall occurs.

3.2.1.3 Coal and gas outburst

Segment OA: damage deceleration increased stage.

Segment AB: stable development of damage stage.

Segment BC: damage acceleration stage.

Segment CD: disaster development stage.

3.2.2 Mechanical accident

Segment OA: run into the wear section of initial parts.

Segment AB: constant speed wear and aging section.

Segment BC: accelerated wear and tear aging section related to component life.

Point C: the critical wear and aging of the machine.

3.2.3 Social change or reform

The vertical axis indicates the intensity of contradiction.

Segment OA: the stage of gradual relaxation of contradictions during the establishment of a new order.

Segment AB: the stable development stage of contradictions after the establishment of the new order.

Segment BC: the contradiction aggravated after the old management system hindered the development of productive forces.

Point C: a sudden change in the degree of conflict intensification.

3.2.4 Human life process

3.2.4.1 Human life process

The vertical axis represents the aging degree of the human body.

Segment OA: from birth to puberty.

Segment AB: puberty to old age.

Segment BC: the accelerated aging phase of the human body entering old age.

Point C: the critical value of life decay.

3.2.4.2 Human casualties

The vertical axis represents the probability of human casualties.

Segment OA: the process of the human body entering the work environment to recognize the work object.

Segment AB: the stage where the probability of casualties increases steadily after familiarizing with the environment.

Segment BC: some objectively lead to increased risk stage.

Point C: critical value of risk.

3.2.5 Safety rheology-mutation theory

In summary, the contradictory movement of safety-danger runs through the existence of things. And it is restricted by internal and external factors. To study the rheology-mutation movement of things, it is necessary to find the internal factors from the risk factors, and obtain the basic law of safety rheology-mutation through studying the changing laws of the internal factors, and then consider the influence of external factors.

The whole process of "safe flow and mutation" of things is:

In the initial stage after the birth of a new thing (segment OA), its damage increases gradually with time, and a new order is gradually formed and perfected during this period.

When the new order develops to the mature stage (segment AB), the perfect new order causes the damage to increase slowly and uniformly.

After a period of steady increase, the original order will again develop in the direction of disorder, and the damage value will start to increase at an accelerated rate (segment BC).

Everything has its inherent damage tolerance limit. Beyond this limit, things will undergo safety mutations.

When the original order is broken, things begin to return to a new safe state (point O).

3.3 The basic theory of safety rheology-mutation

3.3.1 Rheology-mutation basic concept in Safety Science

3.3.1.1 State concept of Safety Science

(1) Safety: the state in which the main body of a thing is damaged within a certain limit, and the limit can be determined according to the safety requirements of specific things.

(2) Danger: the state where the main body of something is damaged beyond a certain limit.

(3) Accidents: the partial or overall order disorder and disintegration of the original thing when the damage amount of the thing reaches the limit value. The accident state in nature is called a disaster.

3.3.1.2 The essential concept of Safety Science

(1) Damage: the amount of damage over time under the action of internal and external factors.

(2) Safety degree: describes the probability value of things maintaining a safe state, $S=1-S_b$, S_b represents the probability that things may have an accident state, which is the degree of danger.

(3) Lifetime limit: the time when the damage reaches the limit under the action of internal factors, or the time it takes for a thing to transition from its birth through an ideal environment to the occurrence of an accident.

(4) System: the object of safety status research is a collection of one or more things.

3.3.1.3 Process concept of Safety Science

(1) Order: the composition, structure and internal operation laws of things themselves.

(2) Safety mutation: the qualitative change process of things from a dangerous state to an accident state.

(3) Safety rheology: the quantitative process of the damage of things over time.

(4) Safety damage (denoted by e): the amount of damage that things destroy over time under the action of internal and external factors.

(5) Safety external factors: external factors that affect the degree of safety of things, refer to all the collective cooperation around things that have an impact on the things themselves.

(6) Safety internal factors: internal factors that affect the degree of safety of things, refer to the interaction and mutual influence of the internal structure, composition, and form of things.

(7) Safety performance: determined by the state of things, reflects the possibility of changes in the movement of things, and is a public measure of the qualitative change and deformation of things in safety.

(8) Safety Entropy: a measure that promotes the transformation of things in the direction of negative effects.

(9) Safety function: it is a characteristic parameter of the process, reflecting the change of the state of things measure.

(10) Hysteresis effect of damage: when the energy used for things is used to damage things, the essential damage will not appear immediately, but an elastic front effect will be produced. When the energy disappears, the damage will not stop immediately, but will be delayed for a period of time. These two times are important for safety prediction.

(11) Safety potential: it is an possible expression that refers to the ability and trend of things to transform from one state to another. And it refers to the ability and trend of things to achieve the best or ideal state of safety.

(12) Potential damage potential: refers to the reflection of the ability and trend damage degree of things to reach the safe best state in the continuous development of things.

3.3.2 Rheology-mutation theoretical model in Safety Science

Physical model: based on a qualitative understanding of the safe rheological-mutation characteristics of things, the following establishes a physical model of safe rheological-mutation. The physical model and model diagram are shown in Figure 3-1 and Figure 3-2, respectively.

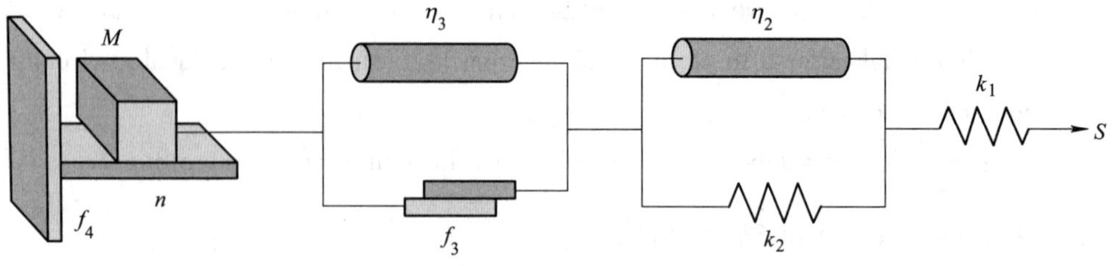

Figure 3-1 The physical model

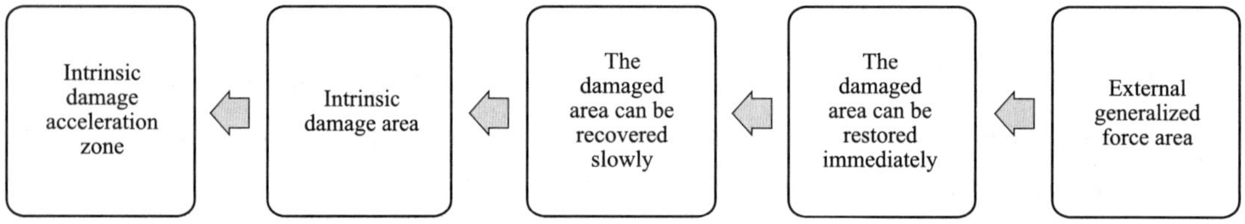

Figure 3-2 Model diagram

3.3.2.1 Four basic elements

(1) Safety reversible element;
(2) Safety damping element;
(3) Safety friction part;
(4) Safety mass body.

3.3.2.2 Five levels

A External generalized force area *S*

The general term for all external influences on things. It acts on things through the inside of things.

When the external change range of the same thing is not large, it can be considered as the same force effect. For example, when studying the law of human life, the external environment is relatively stable, and the physiological and psychological changes are not large. So it can be analyzed under constant force.

B The damaged area can be restored immediately

The first protection zone, which is composed of a safety reversible element, can immediately react to external effects and store the effect in the form of recoverable damage. Once the external effects disappear, the danger to things disappears immediately.

C The damaged area can be recovered slowly

The second protection zone is composed of two components: safety damping element and safety reversible element. Its characteristic is that the force cannot immediately cause due damage. There is a time lag. When the external force disappears, the loss cannot be recovered immediately. But after a period of time. The loss slowly returns to the original position.

D Intrinsic damage area

The irreparable damage area inside the thing is composed of safety damping element and safety friction part. When the force transmitted to the essential area is small, the safety friction part is equivalent to a strength element that protects things and resists external forces without causing essential damage to things. When the force transmitted to the essential area is large, the safety friction part will consume part of the external force, and the remaining force is applied to the safety damping element to form essential damage.

E Intrinsic damage acceleration zone

The essential damage acceleration zone is composed of mass elements, which describe the acceleration of the damage of things. If the external effect exceeds a certain value, it will cause the internal essential damage to accelerate the movement of the component. It can digest or absorb part of external effects. At the beginning, its digestion or absorption capacity is a certain value, but after a period of operation, the quality of the safety mass body will continue to decrease with time, which is a monotonically decreasing function of damage. Even if the external forces remain the same, the damage caused by the mass body will also accelerate. Due to the continuous decrease of the mass body, the ability to protect things from accelerated damage is gradually reduced. A large number of external forces act on the essential damage acceleration zone of the thing. The damage speed of the thing is getting faster and the damage degree is getting bigger and bigger until things are completely destroyed.

In daily life, many safety problems can be simplified into the model diagram of two groups of elements, that is, the second and fourth groups of elements together constitute a safe rheological-mutation model of things. The first and third groups of elements have a relatively stable relationship with time and can be considered as constants. Their graphs are shown in Figure 3-3. It can be clearly seen from the figure that only the second and fourth groups of elements can describe the rheumial state of things.

3.3.2.3 The summary of this chapter

(1) The development history of Safety Science.

(2) Philosophical characteristics of Safety Science.

(3) Application of Boolean algebra, probability calculation and reliability in Safety Engineering.

(4) Rheology-mutation theoretical model in Safety Science, and use it to explain the disaster process of accident.

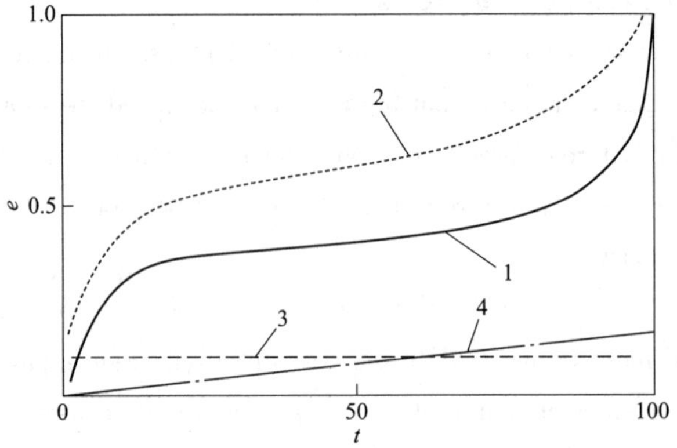

Figure 3-3 Safety rheology-mutation diagram

1—$e_1+e_2+e_3+e_4$;2—e_2+e_4;3—e_1;4—e_3

3.3.3 Modern rheology-mutation theoretical model application in Safety Science

Professor Fu Gui (borns in 1961), professor Jin Longzhe (borns in 1963.08) and professor He Xueqiu (borns in 1961.08) made a lot of contribution to the improvement of the safety theories. They are ones of the first national certified certified safety engineers. Professor Fu Gui advocates taking "the common causes of accidents" as the basis, adopting comprehensive strategies to prevent various accidents, developing accident prevention management tools and practical technologies. Professor He Xueqiu established and discussed systematically "the rheology-mutation mechanism of the occurrence process of coal-rock dynamic disaste" on the basis of a large number of laboratory and field experiments, and depth theoretical research of rheological mechanics, physical chemistry and electromagnetic dynamics.

4 Summary of the Accident

4.1 Definition and characteristics of accidents

4.1.1 Definition of accident

Accidents are Unwanted events that occur in people's production and life activities. Various definitions have been given to accidents, among which the definition of Berckhoff is more famous. According to Burkhoff's definition, an accident is an event that occurs suddenly, violates the will of the person, and forces the activity to stop temporarily or permanently in the course of activities carried out by people (individual or collective) to achieve a certain intention. This definition provides a comprehensive description of the accident.

4.1.1.1 The concept of accident background

Accident background: "The process of activities carried out to achieve a certain intention." An accident is a special event that occurs in human production and life activities. Accidents may occur during any production and life activities of human beings. People must take measures to prevent accidents if they want to carry out activities according to their own intentions.

4.1.1.2 The concept of accident

The definition of an accident: "Suddenly, contrary to human will." An accident happened suddenly and is not expected. This is because the causes of accidents are very complicated and are often caused by many accidental factors, so the occurrence of accidents is random in nature. Before an accident occurs, people cannot accurately predict when, where, and what kind of accident will happen. Due to the random nature of accidents, it is very difficult to recognize accidents, understand the rules of accidents, and prevent accidents.

4.1.1.3 The concept of accident consequences

Consequences of the accident: "Forcing the activity to stop temporarily or permanently". An accident is an event that forces the ongoing production and living activities to stop temporarily or permanently. The interruption of accidents and the termination of activities will inevitably bring about some form of impact on people's production and life. Therefore, an accident is an event against people's will, an event that people do not want to happen.

4.1.2 The main influencing factors of the accident

4.1.2.1 Human

The so-called human factors include operating workers, management cadres, people present at the scene of the accident, and related personnel.

Human unsafe behavior is an important cause of accidents, including:

(1) Operate without permission, ignore safety and ignore warnings;

(2) Dangerous work or high-speed operation;

(3) Artificially invalidate the safety device;

(4) Use unsafe equipment, use hands instead of tools to carry out operation or illegal operations;

(5) Unsafe loading, stacking, and combination of objects;

(6) Adopt unsafe working posture or orientation;

(7) Work on equipment with dangerous operation or on moving equipment work; do not stop the machine, work while overhauling;

(8) Distraction, playfulness, intimidation, etc.

4.1.2.2 Sub factor

The so-called substance factors include raw materials, fuels, power, equipment, tools, finished products, semi-finished products and so on. The unsafe states of things are as follows:

(1) Poor structure of equipment and devices, insufficient material strength, wear and aging of parts and components;

(2) There are dangerous and harmful substances;

(3) The workplace is small or has other defects;

(4) The safety protection device fails;

(5) Lack of protective equipment and clothing or defects;

(6) The stacking and sorting of materials are defective;

(7) The process is unreasonable and the operation method is unsafe.

The unsafe state of things constitutes the material basis of the accident. Without the unsafe state of things, accidents are impossible. The unsafe state of the material constitutes a hidden danger and source of danger in production. When it meets certain conditions, it turns into an accident.

4.1.2.3 Environment factor

An unsafe environment is the material basis for accidents. It is the direct cause of the accident and usually refers to:

(1) Abnormalities in the natural environment, that is, severe variations in rocks, geology, hydrology, meteorology, etc.

(2) Poor production environment, that is, defects in lighting, temperature, humidity, ventilation,

lighting, noise, vibration, air quality, color, etc.

4.1.2.4 Management

The reason for management is the defect of management. it has:

(1) Technical defects refer to defects in the design, material selection, installation arrangement, maintenance and repair of industrial buildings, structures, mechanical equipment, instrumentation, etc.; or there are problems with process flow and operation methods;

(2) Unreasonable labor organization;

(3) Lack of inspection guidance for on-site work, or wrong inspection guidance;

(4) There are no safety operating procedures or are not sound, misappropriation of safety measures costs, failure to implement accident prevention measures seriously, and ineffective rectification of potential safety hazards;

(5) Insufficient education and training, so staff do not understand operating technology or experience, and lack safety knowledge;

(6) Improper selection and use of personnel. Some people have physical or physical defects, such as diseases, poor hearing and vision, etc.

Management deficiencies are the indirect cause of the accident and the condition for the direct cause of the accident to exist.

4.1.2.5 Accident judgment

The handling of the accident refers to:

(1) Whether the correct judgment and response can be made to the abnormal signs before the accident;

(2) Once an accident occurs, whether effective measures can be taken quickly to prevent the situation from deteriorating and expanding the accident;

(3) Whether rescue measures and first aid measures for injured persons are proper.

Obviously, these factors play a restrictive role in the occurrence and development of accidents, and they appear during the accident.

4.1.3 Characteristics of the accident

Characteristics of the accident: causality; chance; necessity; regularity; potential; reproducibility; predictability; preventable.

Standard for classification of accident grades:

(1) According to human injury;

(2) To casualties of enterprise employees;

(3) To consequences of the accident;

(4) To accident severity;

(5) To responsibilities of the accident;

(6) To other standards;

4.2 Accidents classification

According to the Classification for Casualty Accidents of Enterprise Staff and Workers (GB 6441—86), it is divided into: (1) Object strike; (2) Vehicle injury; (3) Mechanical damage; (4) Lifting injury; (5) Electric shock; (6) Drowning; (7) Burning hot; (8) Fire; (9) Falling from high places; (10) Collapse; (11) Take off the top film to help; (12) Permeable; (13) Blasting; (14) Gunpowder exploded; (15) Gas explosion; (16) The boiler exploded; (17) The container exploded; (18) Other explosions; (19) Poisoning and suffocation; (20) Other injuries.

(1) Object strike: personal injury accident caused by the inertial force of an out-of-control object. It is suitable for damage caused by falling objects, flying objects, rolling rocks, and collapsed blocks, but does not include object blows caused by explosions.

(2) Vehicle injuries: mechanical injury accidents caused by motor vehicles. It is suitable for accidents such as squeezing, crushing, crashing or overturning of motor vehicles while driving, getting on and off the vehicle, riding on a mine cart or flying car, vehicle transportation accidents, and sports car accidents. Motor vehicles here refer to: cars, battery cars, tractors, rail cars, excavators, bulldozers, electric shovels, etc.

(3) Mechanical injuries: twisting, rolling, bumping, cutting, poking, cutting and other injuries causedby mechanical equipment and tools. Such as the work piece or tool flying out and hurting people, cutting chips hurting people, the hand or body involved in the machine, the hand or other parts injured by the tool, the rotating machinery wrapping and pressing people, etc. Except in the case of vehicles and lifting equipment.

(4) Lifting injuries: mechanical injuries caused by lifting operations. It is suitable for various lifting operations. Such accidents mainly include: bridge type cranes, boom type cranes, lifts, light and small lifting equipment and other operations. The main types of lifting injuries include unhooking and hitting people during lifting operations, wire rope breaking and drawing people, moving hanging objects hitting people, twisting wire ropes or pulleys, etc. It also includes accidents such as lifting equipment tipping during use and installation accidents, over-rolling of lifting equipment and squatting tank. However, it does not apply to the following injuries: electric shock, injuries caused by brake failure during maintenance, and falls caused by getting up and down the cab.

(5) Electric shock injuries: an accident in which electric current flows through the human body causing physical injury. Electric shock accidents are divided into two categories: electric shock and electric injury. This type of injury accidents mainly includes electric shock and lightning injury: such as human contact with the metal shell of live equipment, exposed temporary wires, and leakage hand-held electric tools; incorrect operation of lifting equipment contacting high-voltage

lines or induction of electrification; lightning injury; electric shock and falling, etc.

(6) Drowning: an accident in which a large amount of water enters the lungs through the mouth and nose, causing obstruction of the respiratory tract and suffocation due to acute hypoxia. This type of injury accidents applies to accidents that occur when ships, rafts, and facilities are sailing, berthing, and operating.

(7) Burns: chemical burns caused by strong acid, strong alkali and other substances splashed on the body; burns caused by flames; scalds caused by high-temperature objects; skin damage caused by radiation. Burns mainly include burns, scalds, chemical burns, radiation skin injuries, etc. But it does not include electric burns and burns caused by fire accidents.

(8) Fire: an enterprise fire accident that caused personal injury or death. This type of accident does not apply to fires caused by non-enterprise reasons. For example, a residential fire spreads to an enterprise. Such accidents are those counted by the fire department.

(9) Falling from high altitude: an injury accident caused by a dangerous gravitational potential energy difference. It is customary to call the work site more than 2m above the ground as high operations, and high-altitude operations generally refer to heights above 10m. This type of accident is suitable for falling above the ground such as scaffolding, platform, steep wall construction, etc. It is also suitable for falling from the ground into a hole, pit, ditch, elevating mouth, funnel, etc. However, it is necessary to exclude falls due to other types of induced conditions. For example, when working at a height, a fall due to an electric shock should be regarded as an electric shock accident and cannot be classified as a fall from high-altitude operations.

(10) Collapse: accidents caused by the collapse of buildings, structures, piles, etc. and the collapse of earth and rock. This type of accident is suitable for collapses caused by unreasonable design or construction, as well as collapses of earth and rocks, such as building collapses, scaffold collapses, soil and rock collapses caused by excavation of trenches, pits, and holes, etc. This type of accident is not applicable to collapse accidents caused by roof fall in mines or explosions or blasting.

(11) Roof fall slab: the collapse of the working face and sidewall of the roadway due to improper support and excessive pressure is called slab; the collapse of the roof is the roof fall. The two often happen at the same time, referred to as the roof and side fall accident. This type of accident is applicable to collapse accidents in mines, underground mining, tunneling and other tunnel operations.

(12) Water penetration: casualties caused by accidental water sources in mines, underground mining or other tunnel operations. This type of accident is suitable for accidents occurs when the well is connected to aquifer rock formations, underground aquifers, karst caves, or flooded roadways and surface waters. Water gushes into disaster. It is not applicable to surface water damage accidents.

(13) Shooting: casualties caused by shooting during construction. This type of accident is

suitable for various blasting operations, such as casualties caused by blasting operations in quarrying, mining, coal mining, mountain driving, road repairing, and demolition of buildings.

(14) Gunpowder explosion: explosion accidents that occur during the production, transportation and storage of gunpowder and explosives. This type of accident applies to chemical explosion accidents, caused by vibration, open flame, friction, static electricity, or the thermal decomposition of explosives, the storage time is too long, or the storage volume is too large when gunpowder and explosives are produced during the batching, transportation, storage, and processing processes. And when smelting metals, waste disposal is not clean, and the remaining gunpowder or explosives cause explosion accidents.

(15) Gas explosion: a chemical explosion accident caused by the mixture of combustible gas, coal dust and air with a concentration reaching the combustion limit, contacting the ignition source. This type of accident is suitable for coal mines, but also for occasions where the air does not circulate and gas and coal dust accumulate.

(16) Boiler explosion: use various fuels, electricity or other energy sources to heat the contained liquid to certain parameters and carry a certain pressure. The scope is specified as a pressurized steam boiler with a volume greater than or equal to 30L; Pressurized hot water boilers with outlet water pressure greater than or equal to 0. 1MPa (gauge pressure) and rated power greater than or equal to 0. 1MW; physical explosion accidents in organic heat carrier boilers. But it does not apply to boilers on railway locomotives, ships, and boilers for train and ship power stations.

(17) Explosion of pressure vessel: a pressure vessel refers to a closed device that contains gas or liquid and carries a certain pressure. Its range is specified as the maximum working pressure is greater than or equal to 0. 1MPa (gauge pressure). Gas, liquefied gas with the product of pressure and volume is greater than or equal to 2. 5MPa/L, fixed container and mobile container of liquid with the highest working temperature higher than or equal to the standard boiling point. Gas, liquefied gas with the nominal working pressure greater than or equal to 0. 2MPa (gauge pressure). And gas cylinders, oxygen chambers, etc. with the product of pressure and volume greater than or equal to 1. 0MPa/L. Liquefied gas, and liquid whose standard boiling point is equal to or lower than 60℃.

(18) Other explosions: explosions that are not classified as gas explosions, boiler explosions and pressure vessel explosions. It mainly includes explosions caused by explosive gases formed by mixing combustible gas with air, or mixing combustible vapor with air or combustible dust mixed with air.

(19) Poisoning and suffocation: poisoning refers to human contact with toxic substances, and the general term for various physiological phenomena that appear in the human body within 8 hours such as eating poisonous food by accident, breathing toxic gas. It is also called acute poisoning. Asphyxiation refers to Abandoned tunnels, shafts, culverts, underground pipes and other places that cannot be ventilated, Because of lack of oxygen, sometimes sudden fainting and even

death accidents occur. The two phenomena merge into one, called poisoning and suffocation accidents. This type of accident neither applies to poisoning and suffocation accidents caused by pathological changes, nor apply to deaths caused by chronic poisoning occupational diseases.

(20) Other injuries: all accidents that do not belong to the above injuries are referred to as other injuries, such as sprains, falls, frostbite, beast bites, nail stick injuries, etc.

4.2.1 Classification of injury degree

China's *Classification of Accidents for Enterprise Employees' Injuries and Deaths* is classified according to the degree to which the injury caused to the injured leading to the loss of work ability. The injuries of casualties are classified into minor injuries, severe injuries and deaths.

(1) Minor injuries refer to disability injuries where the lost working day is more than 1 working day and less than 105 working days.

(2) Serious injury refers to the loss of working days equivalent to 105 working days or more than 105 working days which is determined in Appendix B of the Classification of Accidents for Enterprise Employees' Injuries and Deuths, and the maximum lost working days of serious injuries does not exceed 6000 working days.

There are certain rules for the determination of serious injuries in the daily management of casualties. Any one of the following situations shall be treated as serious injuries: 1) Disabled or likely to be disabled diagnosed by the doctor; 2) The injury is serious and requires major surgery to recover; 3) Important parts of the human body are severely burned or scalded; 4) Severe fractures (sternum, ribs, spine bones, clavicles, shoulder blades, wrist bones, leg bones and foot bones), severe brain oscillation; 5) Eye injuries are severe, and may be blindness; 6) Hand injuries (1 section of thumb, or 2 section of index finger, middle finger, ring finger, little finger or 1 section of any two are broken, local tendon may be disabled); 7) Foot injuries; 8) Internal injuries (internal organ damage, internal bleeding or injury to the peritoneum); 9) Other injuries.

(3) Death refers to 6000 lost working days. This is calculated based on the average retirement age and average age of death of Chinese employees.

4.2.2 Classification of injury

The classification of accidents here is calculated according to the lost working days caused by casualties, that is, the working days when the injured person loses the ability to work (abbreviated as disability).

According to the provisions of the *Classification of Accidents for Enterprise Emplayees' Injuries and Denths*, the inability to work can be divided into the following three situations.

(1) Permanent total incapacity refers to no death, but makes it impossible for the injured person to engage in a remunerated occupation, or an accident causes any one of the following three situations to be disabled (or not disabled but complete loss of function): two eyes; one eye and one

hand(arms, feet, legs) ; any two of the following limbs that are not on the same side: hands, arms, feet, legs.

(2) Permanent partial incapacity refers to a limb or a part of a limb that is mutilated or out of function, or that the whole body or part of its function has been permanently damaged, regardless of the condition of the limb or body function before the injury.

(3) Temporary incapacity refers to the inability to work for one or more days without causing death or permanent injury. Temporary incapacity can be divided into two types: temporary partial incapacity and temporary total incapacity. The working days lost due to temporary disability are calculated based on the actual number of days off work. The day of injury is not counted, but it includes all calendar days(including Saturdays, Sundays, and holidays, etc.) from then to before returning to work.

4.2.3 Classification of accident severity

According to the severity of injuries suffered by employees and the number of injuries, accidents are divided into the following types.

(1) Minor injury accident: an accident with only minor injuries. It refers to an accident that causes the injured person to lose one or more working days without reaching the level of serious injury.

(2) Serious injury accidents: there are only serious injuries but no deaths in the accident. It causes serious injuries to the operator's brain, eyes, limbs, torso and other parts, which leads to accidents where the ability to work is lost.

(3) Fatal accident: 1 to 2 people died in an accident.

(4) Major fatal accidents: 3 to 9 people died in an accident.

(5) Extraordinary fatal accident: 10 to 29 people(including 10 people) died in an accident.

(6) Extraordinarily serious fatal accidents: more than 30 people(including 30 people) died in an accident.

The above classification method classifies accidents with human casualties mainly based on the severity of injuries and forgets, without considering the economic losses caused by the accident. In order to comprehensively consider the losses caused by accidents (including casualties and economic losses) and strengthen accident management, according to the relevant provisions of Article 3 of the *Regulations on Reporting and Investigation and Handling of Production Safety Accidents*, production accidents are generally divided into the following four levels:

(1) A particularly serious accident: refers to an accident that caused more than 30 deaths; or more than 100 serious injuries(including acute industrial poisoning, the same below) ; or direct economic losses of more than 100 million yuan.

(2) Heavy accident: refers to an accident that caused more than 10 deaths and less than 30 deaths; or serious injuries of more than 50 and less than 100; or direct economic losses between 50

million and 100 million yuan.

(3) Major accidents: refer to accidents that cause more than 3 deaths and less than 10 deaths; or serious injuries of more than 10 people and less than 50 people; or direct economic losses between 10 million and 50 million.

(4) General accident: it refers to an accident that causes the death of less than 3 people; or serious injury of less than 10 people; or direct economic loss of less than 10 million yuan.

4. 2. 4 Other categories of accidents

(1) According to the different responsibilities for causing the accident, it can be divided into two categories: responsible accidents and non-responsible accidents. Responsible accidents are caused by people violating laws, regulations, rules, and standards. Non-responsible accidents are caused by irresistible natural factors or causes that cannot be predicted by current science.

(2) According to the different consequences of the accident, it can be divided into casualty accidents and non-injury accidents. Casualty accidents cause personal injury. Non-injury accidents only cause production interruption, equipment damage or property loss.

(3) According to the different industries of accident supervision and management, it can be divided into enterprise employee casualties (industrial, mining and commercial enterprises), fire accidents, road traffic accidents, water traffic accidents, railway traffic accidents, civil aviation flight accidents, agricultural machinery accidents, fishing vessels Accidents and other accidents.

(4) In addition, accidents can also be classified according to the location of the injury, the nature of the injury, and the factors causing the injury.

4. 3 Accident statistics

4. 3. 1 Contents of accident statistics

The contents of accident statistics include employee casualty accident registration form, accident investigation report and approval or accident handling decision, on-site investigation records, drawings, photos, etc., as well as technical appraisal, inspection conclusions and experimental reports, witnesses, physical evidence materials, direct economic loss materials, medical treatment The unit's medical certificate (certificate) for casualties, technological conditions, operating conditions and design data of the accident, notices, briefings and documents related to the accident. The contents of the accident unit that need to be counted include the unit name, unit address, unit mailing address, unit code, postal code, contact number, number of employees, enterprise scale, economic type, industry, competent department, etc. A limited liability company is a company registered in accordance with the *Regulations of the People's Republic of China on the Administration of Company Registration* and is jointly invested by two or more but less than 50 shareholders. An economic organization that is liable for its liabilities with all its assets. A company limited by shares refers to a company registered in accordance with the *Regulations of the People's*

Republic of China on the Administration of Company Registration. Its entire registered capital consists of equal shares and capital is raised by issuing shares. An economic organization responsible for its debts. A private enterprise refers to a profit-making economic organization invested and established by a natural person or controlled by a natural person and based on hired labor, including private limited liability companies registered in accordance with the provisions of *the Company Law*, *the Partnership Enterprise Law*, and *the Provisional Regulations on Private Enterprises*. Corporations, private limited companies, private partnerships and privately owned sole proprietorships. Hong Kong, Macao and Taiwan investment enterprises include joint ventures (Hong Kong or Macao, Taiwan investment), cooperative enterprises (Hong Kong or Macao, Taiwan investment), Hong Kong, Macao and Taiwan sole proprietorship enterprises, Hong Kong, Macao and Taiwan investment shares limited liability company. Foreign-invested enterprises include Sino-foreign joint ventures, Sino-foreign cooperative ventures, foreign-funded enterprises, and foreign-invested joint-stock companies.

Other enterprises refer to other economic organizations other than those mentioned in Articles 1 to 9 above.

4.3.2 Accident situation

The contents of the accident situation that need to be counted include the place of occurrence, the date of occurrence, the time of occurrence, the type of accident, the total number of casualties, the casualties of non-enterprise personnel, the cause of the accident, the victim's loss of working days, the direct economic loss, whether the case was closed, the cause, and the cause of the accident. Harm, unsafe state, unsafe behavior.

4.3.3 Personnel situation

The contents that need to be counted include name, gender, age, type of work, length of service, education level, professional title, injured part, degree of injury, nature of injury, type of employment, date of death, downtime, resumption of work on the date of death, lost working days, and compensation amount.

4.3.4 Situation of coal mining enterprises

The contents of coal mining enterprises, classification of accident locations, accident categories are need to be counted. The contents of coal mining enterprises include the economic type of coal mines, and the economic types of coal mines include state-owned (original state-owned key, former state-owned local) coal mining enterprises, collective coal mining enterprises, and private coal mining enterprises. The accident categories include raw coal production, non-state Raw coal production, capital construction, causes of damage, types of accidents in coal mining enterprises, classification of coal mining types, mortality rate per million tons.

This part introduces four kinds of accident statistical index systems, namely comprehensive casualty accident statistical index system, industrial and mining enterprise casualty accident statistical index system, industry statistical index system, and regional safety evaluation statistical index system. In August 1948, the International Labor Organization meeting established statistical indicators of casualty accidents based on the frequency of casualty accidents and the rate of injury severity . The formula for calculating the frequency of casualties is as follows:

$$a = \frac{A}{NT} \tag{4-1}$$

where, a is the frequency of casualty accidents; A is the number of casualty accidents; N is the number of employees participating in production; and T is the statistical period.

China stipulates that the frequency of casualty accidents is calculated according to the death rate of 1000 people, the serious injury rate of 1000 people and the frequency of injuries. The 1000-person mortality rate is the number of deaths due to work-related accidents per 1000 employees in a certain period .

$$\text{Death rate per thousand} = \frac{\text{Death toll}}{\text{Average number of employees}} \times 10^3 \tag{4-2}$$

The 1000-person serious injury rate refers to the average number of people seriously injured by work-related accidents per 1000 employees in a certain period of time.

$$\text{1000 serious injury rate} = \frac{\text{Seriously injured}}{\text{Average number of employees}} \times 10^3 \tag{4-3}$$

Injury frequency refers to the average number of injuries caused by work-related accidents per million working hours in a certain period.

$$\text{Damage frequency} = \frac{\text{Number of people injured}}{\text{Actual total hours}} \times 10^6 \tag{4-4}$$

Note: At present, China still uses the frequency of work-related accidents stipulated by the labor department as a statistical indicator, and it is customary to call it the injury rate per thousand people.

$$\begin{aligned}&\text{Frequency of work-related accidents}\\ &= \frac{\text{Number of work-related accidents during this period}}{\text{Number of registered employees during the period}} \times 10^3\end{aligned} \tag{4-5}$$

The accident severity rate refers to the national standard Classification of Accidents for Enterprise Employees' Injuries and Deaths, and the accident severity rate is calculated according to indicators such as the injury severity rate, the average injury severity rate and the mortality rate calculated by product output.

4.3.5 Production safety accident reporting system

The content to be reported in the production safety accident report system includes the scope of statistics, the content of statistics, the type of report form and the unit that fills in the report, the report submission procedure, and the submission time. The reporting time of the accident is that for a major casualty accident with more than 3 deaths, it must be reported to the relevant state departments within 24 hours . The monthly report on casualty accident statistics must be submitted before the 20th of the following month, and the annual report on casualty accident statistics must be submitted to the State Economic and Trade Commission before November 30th of the following year . The State Coal Mine Safety Supervision Bureau is required to report to the State Economic and Trade Commission before the 20th of the following month and the 30th of January of the following year, respectively, for the monthly and annual reports on the statistics of coal mine casualties . For fire accidents, road traffic accidents and other statistical monthly and annual reports, please send the Ministry of Public Safety and the Ministry of Communications to the relevant state departments before the 20th of the next month and the 30th of November of the following year, and report them to the State Council in a timely manner.

4.3.6 Calculation of economic loss of casualty accidents

The economic loss C_T of the casualty accident can be calculated by the sum of the direct economic loss C_D and the indirect economic loss $C_{I,}$ namely

$$C_T = C_D + C_I \tag{4-6}$$

where, C_D for direct economic loss; C_I for indirect economic loss. Direct economic losses include work losses; expenses incurred for personal injury and death; and aftercare expenses. Expenses for personal injury include medical expenses (including nursing expenses), funeral and pension expenses, subsidies and relief expenses, and lost work expenses, etc. ; after-care expenses: including transaction expenses for handling accidents, on-site rescue expenses, on-site cleaning expenses, accident expenses, etc. Fines and compensation costs; property damage costs: including loss of fixed assets and loss of current assets. The formula for calculating direct economic loss is as follows.

$$L = D\frac{M}{SD_0} \tag{4-7}$$

where, L is the value of work loss; D is the number of lost working days; S is the average number of employees in the previous year; M is the company's profit in the previous year; and D_0 is the legal number of working days in the previous year. The death of an employee is calculated as 6000 working days.

Indirect economic losses include production stoppage, production loss value, work loss value, resource loss value, costs for dealing with environmental pollution, training costs for replenishing new employees, and other loss costs. A calculation method that comprehensively considers tangible and intangible losses.

$$\text{Total accident loss} = \text{economic loss} + \text{non-economic loss} \tag{4-8}$$

$$\text{Non-economic loss} = \text{direct non-economic loss} + \text{indirect non-economic loss} \tag{4-9}$$

The direct non-economic loss of the accident includes: the loss of the value of human life andhealth, and the impact on health can be estimated by the impact of working ability .

$$\text{Health value loss} = (1\text{-}K)dy \qquad (4\text{-}10)$$

In the formula, d is the number of working days after resumption of work until retirement, which can be calculated by the number of working years after resumption of work $\times 250$; K is the coefficient of recovery of healthy body function (K is less than 1), and y is the working day considering the appreciation of the value of the working day. value. The loss of environmental damage shall be calculated according to the cost of environmental pollution treatment and its unrestored environmental value. Indirect noneconomic losses of accidents include work efficiency impact, reputation loss, and loss of political and social stability. The ergonomics impact is due to the accident affecting the psychological state of the employees, resulting in a decrease in work efficiency, which can be calculated by the time efficiency coefficient method:

$$\text{Impact loss} = \text{Impact time work efficiency(output value/day)} \times \text{Impact coefficient} \qquad (4\text{-}11)$$

The influence coefficient is determined by the number of employees involved and the degree of influence, in decimals. Reputation loss can be estimated by the amount of decline in the operating efficiency of the company's products, including the loss of product quality and the impact of accidents on product sales, and can be calculated by the coefficient method .

$$\text{Reputation loss} = \text{Original sales value} \times \text{Accident impact factor} \qquad (4\text{-}12)$$

Political and social stability is a potential loss, which can be estimated as a proportion of the total economic loss of the accident.

4.3.7 Basic methods of accident statistics-statistical overview

Arrangement, analysis, description, and inference methods of social and economic phenomenon data, including the methods of acquiring data and making good use of these data. Statistical work refers to the practice of collecting, sorting, analyzing and providing data as statistics. Statistical data refers to various data obtained in the process of statistical practice activities as statistics. Statistics refers to the methods and theories about collecting, arranging, analyzing and interpreting data as statistics.

4.3.8 Comprehensive index method

The comprehensive index is a statistical index that reflects the characteristics of the accident obtained through investigation, statistics, and sorting. According to the different quantitative characteristics of the overall phenomenon it reflects, it can be divided into total index, relative index and average index.

4.3.8.1 Aggregate indicators

Total index, also known as the statistical absolute number, is a comprehensive index used to reflect the scale and level of the accident development. The total amount index is divided into the overall unit total amount and the overall label total amount according to the difference in reflecting the overall content . According to the different reaction time conditions, it is divided into period indicators and time point indicators. According to the different units of measurement used, it is divided into physical indicators, property indicators and working day indicators.

4.3.8.2 Relative indicators

Relative indicators, also known as statistical relative numbers, are the ratio of the values of two related phenomena to reflect the degree of development, prevalence or proportional relationship of the phenomenon. Relative indicators of structure, relative indicators of proportions, relative indicators of comparison, relative indicators ofstrength.

4.3.8.3 Average metrics

Average index, also known as the statistical average, is used to reflect the general level achieved by a certain unit of the total accident population, and the quantity indicator reaches a certain time and place conditions. Features: The difference of the overall unit flag values is abstracted; the average index is a representative value, representing the general level of the overall unit flag values. Function: reflect the central tendency of the distribution of each unit variable in the population; Compare the development levels of similar phenomena in different units; analyze the dependencies between accident phenomena.

Numerical means include: arithmeticmean, harmonic mean, geometric mean.

Location averages include: mode, median.

4.3.8.4 Variation index

The degree of difference or dispersion of accidents in various systems and places. Function: reflect the deviation trend of the accident unit mark value; explain the representative degree of the average index; measure the balance and stability of the phenomenon change.

4.4 Accident investigation and handling

4.4.1 Incident reporting

After the accident occurs, the relevant personnel at the accident site shall immediately report to the person in charge of the unit; the person in charge of the unit shall report to the safety production supervision and management department of the people's government at or above the county level where the accident occurred within 1 hour after receiving the report. Reporting to relevant departments of management responsibilities. Particularly major accidents and major accidents are reported to the safety production supervision and management department of the State Council and the relevant departments responsible for safety production supervision and management; major accidents are reported to the safety production supervision and management departments of the people's governments of provinces, autonomous regions and municipalities directly under the Central Government . Relevant departments with the responsibilities of work safety supervision and management; general accidents are reported to the work safety supervision and management departments of the people's governments at the districted city level and the relevant departments with work safety supervision and management responsibilities. The safety production supervision and management department and the relevant departments with the responsibility of safety production supervision and management shall report the accident situation level by level, and the reporting time at each level shall not exceed 2 hours. The accident report shall include the following: an overview of the unit where the accident occurred; the time and place of the accident, and the situation at the accident site; a brief history of the accident; Losses; Actions Taken ; Other Reportable Circumstances. If any new situation occurs after the accident report, a supplementary report shall be made in a timely manner . Within 30 days from the date of the accident, any change in the number of casualties caused by the accident shall be reported in a timely manner . Within 7 days from the date of occurrence of a road traffic accident or fire accident, if the number of casualties caused by the accident changes, a supplementary report shall be made in a timely manner.

4.4.2 Accident investigation

4.4.2.1 Preparation stage

(1) Preparation stage 1 is accidents reported. The basic information of the accident, such as the accident unit, accident type, casualty situation, accident time, accident location, and simple occurrence of the accident, should be recorded in detail.

(2) Preparatory stage 2 is rushing to the scene. After the safety production supervision and management department and the relevant departments with the responsibility of safety production supervision and management receive the accident report, the person in charge should immediately rush to the accident scene and organize the accident rescue together with the people's government of the place where the accident occurred.

(3) Preparatory stage 3 is formation of investigation team. The accident investigation team is established according to the procedures stipulated in the *Regulations on Accident Reporting and Investigation and Handling. The Regulations on the Reporting, Investigation and Handling of Production Safety Accidents* (Decree No. 493 of the State Council), which came into effect on June 1, 2007, clearly stated: 1) The State Council or relevant departments authorized by the State Council shall organize an accident investigation team to conduct investigation; Accidents, major accidents, and general accidents shall be investigated by the provincial people's government, the people's government of a city divided into districts, and the people's government at the county level at the place where the accident occurred. Entrust relevant departments to organize accident

investigation teams to conduct investigations; for general accidents that do not cause casualties, the people's government at the county level may also entrust the accident occurrence unit to organize accident investigation teams to conduct investigations. 2) When the people's government at a higher level deems it necessary, it may investigate the accident under the responsibility of the people's government at a lower level. Within 30 days from the date of the accident (within 7 days from the date of the occurrence of road traffic accidents and fire accidents), if the accident level changes due to the change in the number of casualties in the accident, and the people's government at the higher level shall be responsible for the investigation in accordance with the provisions of these Regulations, the higher-level people's government shall be responsible for the investigation . The people's government may organize another accident investigation team to conduct investigation. 3) For an accident below a particularly serious accident level, where the place where the accident occurred and the unit where the accident occurred are not in the same administrative region at or above the county level, the people's government at the place where the accident occurred shall be responsible for the investigation, and the people's government at the place where the accident unit is located shall send personnel to participate. The composition of the accident investigation team should follow the principles of simplicity and efficiency. The members of the accident investigation team shall have the knowledge and expertise required for accident investigation and shall have no direct interest in the accident under investigation. The leader of the accident investigation team shall be designated by the people's government in charge of the accident investigation, and the leader of the accident investigation team shall preside over the work of the accident investigation team. The responsibilities of the accident investigation team are: to find out the process, cause, casualties and direct economic losses of the accident; to determine the nature of the accident and responsibility for the accident; to put forward suggestions for dealing with those responsible for the accident; action; submit incident investigation report.

(4) Preparation Phase 4 is developing an investigation plan. According to the statutory responsibilities of the accident investigation team, an investigation work plan is developed. The investigation group generally consists of three groups: the technical group, the management group, and the comprehensive group.

(5) Preparatory stage 5 is investigation team meeting. The accident investigation team is composed of the relevant people's government, the safety production supervision and management department, the relevant departments with the responsibility of safety production supervision and management, the supervision organ, the public safety organ and the trade union, and should invite the people's procuratorate to send people participate. After the accident investigation team is approved to be established, a meeting of the accident investigation team shall be held to report the accident situation and announce the investigation work plan . The members of each investigation team shall have a clear division of labor and tasks.

4.4.2.2 Investigation phase

(1) Investigation Phase 1 is on-site investigation. Before on-site investigation, the parties or insiders of the accident should be asked to understand the basic information of the accident, such as the time, place, process, and disaster-affected location of the accident. Before the survey, first patrol the entire site, overall picture of the site, determine the scope and sequence of the site survey. Extract relevant traces and physical evidence left at the scene of the accident. According to the on-site investigation and on-site evidence collection, draw relevant drawings of the accident scene. According to the situation of on-site investigation, after drawing relevant drawings, submit the accident site investigation report, indicating the accident site investigation personnel, investigation time, investigation route, and explaining the basic situation of the accident site and the situation related to the accident.

(2) Investigation Phase 2 is data collection. When the accident investigation is carried out, relevant data related to the accident such as the production rules and regulations, safety responsibility system, operation (operation) procedures, and personnel training of the accident unit should be collected.

(3) Investigation Phase 3 is extracting physical evidence. Before and after the extraction, it should be recorded in various forms such as video, photography, and text (with the current field surveys are carried out simultaneously).

(4) Investigation Phase 4 is witness testimony. Find all the escapees at the accident scene as soon as possible, and understand the truth of the accident. The inquiries shall be recorded and signed or stamped by law enforcement officers and persons under investigation. Find the person under investigation related to the accident as soon as possible (the person in charge of the accident unit, the leaders and staff of the local government and its relevant departments) to talk and inquire. After the investigation and inquiries are completed, all investigation and evidence collection records shall be sorted out, and according to the investigation and inquiries, it shall be decided whether to supplement the investigation and inquiries.

(5) Investigation Stage 5 is calculation of direct losses. During accident investigation, it is necessary to collect relevant data and calculate the direct economic loss of the accident.

(6) Investigation Stage 6 is technical appraisal. If technical appraisal is required in accident investigation, the accident investigation team shall entrust a unit with the qualifications specified by the state to carry out technical appraisal. When necessary, the accident investigation team may directly organize experts to conduct technical appraisal. The time required for technical appraisal is not included in the accident investigation period. After surveying the site, investigating personnel involved in the accident, and collecting relevant evidence, a unit with the qualifications specified by the state or an expert directly organized by the investigation team shall issue a technical appraisal report of the accident and analyze the direct cause of the accident.

4.4.2.3 Analysis phase

(1) Analysis Phase 1 is cause analysis. The immediate causes of the accident include the unsafe state of things, the unsafe behavior of people. Indirect causes include analyzing the technical reasons from the technical management point of view from the relevant technical data and evidence such as the drawings and regulations of the accident unit. And the training and education that led to the accident. From the aspect of safety management of the accident unit, it is analyzed to find the indirect cause of the accident that the accident unit neglected the aspect of safety management. The investigation understands and analyzes the physical and mental state of the parties involved in the accident at the time of the accident.

(2) Analysis Phase 2 is accident qualitative. The nature of the accident: it is divided into responsible accidents and non-liability accidents. Accidents caused by human force majeure are non-responsible accidents.

(3) Analysis stage 3 is responsibility analysis. According to the direct cause of the accident, determine the person directly responsible for the accident, list the violations of laws and regulations, and propose handling suggestions.

(4) Analysis stage 4 is prevention measures. According to the accident causes obtained from the analysis, targeted accident prevention measures are formulated. Propose accident prevention measures to the accident unit and relevant local departments to ensure that the recurrence of similar accidents can be effectively prevented.

(5) Analysis stage 5 is notification. Report the investigation situation, analyze the cause of the accident, propose preventive measures, etc. (responsible personnel will be notified according to relevant procedures).

4.4.2.4 Trial stage

(1) Trial stage 1 is discussing and handling opinions. The investigation team holds a meeting of all personnel to negotiate and discuss the handling opinions of the person responsible for the accident and the responsible unit according to the analysis of the cause of the accident and the violation of laws and regulations of the person responsible for the accident.

(2) Trial stage 2 is formation of investigation report. The content of the accident report should include: overview of the accident unit; accident process and accident rescue; casualties and direct economic losses caused by the accident; the cause of the accident and the nature of the accident; Identification and handling suggestions for those responsible for the accident; accident prevention and rectification measures. The accident investigation report shall be accompanied by relevant evidential materials. The members of the accident investigation team shall sign the accident investigation report.

(3) Trial Phase 3 is review and investigation report. The accident leading organization unit

convened a bureau meeting to review the accident investigation report. If any problem is found, supplement the investigation or amend the accident investigation report according to the actual situation.

(4) Trial Stage 4 is data archiving. After the investigation and handling of the accident is completed, the investigation report, technical appraisal report, relevant evidence materials and other related materials will be sorted and archived according to the principle of one volume and one file .

4.4.2.5 Handling stage

(1) Handling Stage 1 is accident reply. For major accidents, major accidents, and general accidents, the people's government in charge of accident investigation shall make an approval within 15 days from the date of receipt of the accident investigation report; for particularly major accidents, make an approval within 30 days, under special circumstances, the approval time can be appropriately extended, but the extended time shall not exceed 30 days.

(2) Handling Stage 2 is implementing the reply opinions. Relevant organs shall handle the matter separately in accordance with the approval of the people's government, and in accordance with the powers and procedures stipulated by laws and administrative regulations: administrative penalties are imposed on the unit where the accident occurred and the relevant personnel . Punish the state staff responsible for the accident. The accident unit shall deal with the personnel responsible for the accident. If the person responsible for the accident is suspected of committing a crime, he shall be investigated for criminal responsibility according to law. The accident unit implements preventive and corrective measures to prevent the accident from happening again.

(3) Handling Stage 3 is announcement of handling information. The handling of the accident shall be announced to the public by the people's government in charge of the accident investigation or the relevant departments or agencies authorized by it, except those that should be kept secret in accordance with the law.

(4) Handling Stage 4 is supervision and inspection. The safety production supervision and management department and the relevant departments with the responsibility of safety production supervision and management shall supervise and inspect the implementation of preventive and rectification measures by the accident unit. The implementation of preventive and corrective measures shall be subject to the supervision of the trade union and employees.

By investigating the responsibility of the accident, the person responsible for the accident can be educated, and the leaders of the accident unit and the majority of employees can learn from it. This will improve their work, improve the level of accident prevention, and also play a warning role for other members of the society. According to the *Regulations on the Reporting, Investigation and Handling of Production Safety Accidents* (Decree No. 493 of the State Council), the legal responsibilities of the accident unit and its relevant responsible persons are as follows. If the main person in charge of the accident unit does not immediately organize the accident rescue, report late or fail to report the event of an accident, or leaving duty without permission during the investigation and handling of the accident, a fine of 40% to 80% of the annual income of the previous year shall be imposed. If the person belongs to a state functionary, he shall be punished according to law. If a crime is constituted, the criminal shall be investigated according to law responsibility.

The unit where the accident occurred and its relevant personnel falsely reported or concealed the accident; forged or deliberately destroyed the accident scene; transferred or concealed funds or property, or destroyed relevant evidence and materials; refused to accept investigation or refused to provide relevant information and materials; Whoever commits perjury or instigates others to perjury during the accident investigation; if one of the acts of escaping and hiding after the accident occurs, a fine of not less than 1 million yuan but not more than 5 million yuan shall be imposed on the unit where the accident occurred; and other directly responsible persons shall be fined 60% to 100% of the previous year's annual income; those who belong to the state functionaries shall be punished according to law; those who constitute violations of public safety management shall be punished by public safety organs according to law ; If a crime is committed, criminal responsibility shall be investigated according to law:

The accident unit is responsible for the accident, it shall be fined in accordance with the following provisions: if a general accident occurs, a fine of not less than 100 000 yuan but not more than 200 000 yuan shall be imposed; in the case of a major accident, a fine of not less than 200 000 yuan but not more than 500 000 yuan shall be imposed. If a major accident occurs, a fine of not less than 500 000 yuan but not more than 2 million yuan will be imposed; if a particularly major accident occurs, a fine of not less than 2 million yuan but not more than 5 million yuan will be imposed.

If the main person in charge of the unit where the accident occurred fails to perform the duties of production safety management according to the law, causing the accident to occur, a fine shall be imposed in accordance with the following provisions; if it is a state functionary, it shall be punished according to law; if a crime is constituted, criminal responsibility shall be investigated according to law: if a general accident occurs If a major accident occurs, a fine of 40% of the previous year's annual income will be imposed; if a major accident occurs, a fine of 60% of the previous year's annual income will be imposed; if a particularly serious accident occurs Yes, a fine of 80% of the previous year's annual income will be imposed.

If the accident unit is responsible for the accident, the relevant department shall temporarily suspend or revoke its relevant license according to law. The relevant personnel who are responsible for the accident at the accident unit shall have their professional qualifications related to production safety suspended or revoked according to law. Post certificate: if the main person in charge of the unit where the accident occurred is subject to criminal punishment or dismissal, he shall not serve as the main person in charge of any production and business unit within 5 years from the date of completion of the execution of the punishment or the date of the punishment. For intermediaries that provide false certificates for the units in which the accident occurred, the relevant departments shall temporarily suspend or revoke their relevant certificates and licenses and the practicing qualifications of their relevant personnel. If a crime is constituted, criminal responsibility shall be investigated according to law.

5 Accident Causation Theory and Model

5.1 The emergence and development of accident cause theory

The basic theory of safety science is the accident mechanism and accident model extracted from the analysis of the essential causes of a large number of typical accidents. Therefore, the accident cause theory or accident model can also be called the basic theory of safety science.

The accident cause model reflects the regularity of accidents, can qualitatively and quantitatively analyze the causes of accidents, and provides a scientific and complete theoretical basis for accident prevention and human safety behavior. The accident models effectin production safefy is demonstrated in Figure 5-1.

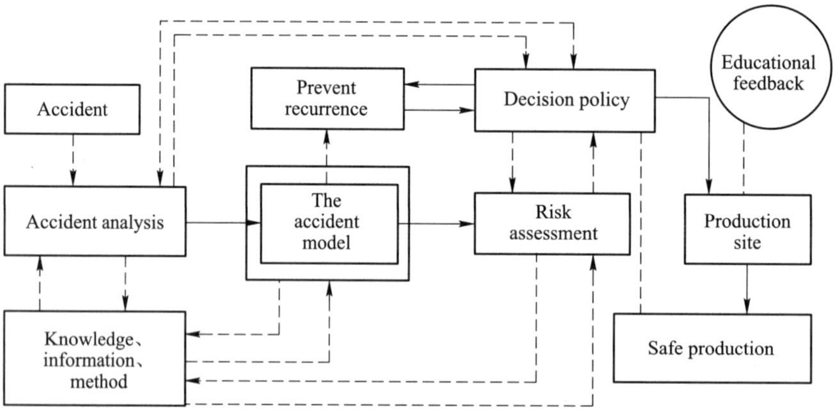

Figure 5-1 The accident models effect in production safety

With the development of science and technology and production methods, the essential laws of accidents are constantly changing, and people's understanding of the causes of accidents is also deepening. Therefore, more than a dozen representative accident cause theories and accident models have appeared.

Early accident cause theory: after WWII accident cause theory, modernaccident cause theory and model.

5.2 Accident proneness

5.2.1 Overview of accident proneness

Accident Proneness: Individuals are prone to accidents, stable, personal inherent tendency.

In 1919, Greenwood and Woods conducted statistical tests on the number of injuries in many factories according to the following three statistical distributions:

(1) Poisson distribution: when there is no individual difference in the probability of accidents, that is, when there is no tendency for frequent accidents, the number of accidents in a certain period of time follows the Poisson distribution. In this case, the accident occurred due to the production conditions in the factory, mechanical equipment problems, and some other accidental factors.

(2) Biased distribution: some workers have mental or psychological problems. If an accident occurs during the production operation, it will cause timidity or nervousness. When the operation is continued, the second and third accidents will happen repeatedly. tendency. This statistical distribution is just based on the few who have mental or psychological defects.

(3) Distribution of unequal liability: when there are many people who are particularly prone to accidents in the factory, the number of people who have different times of accidents obeys non-equal distribution, that is, the probability of accidents is different for each person. In this case, the accident occurred mainly due to human factors.

In order to test the stability of the tendency of frequent accidents, they also calculated the correlation coefficient of the number of accidents that occurred in the first three months and the next three months by the same person in the factory under investigation. The results found that there are accident-prone tendencies in the factory, and the correlation coefficient of the number of accidents in the first and last three months varies from 0.37 ± 0.12 to 0.72 ± 0.07, which are all positively correlated.

In 1926, E. M. Newbold studied the distribution of the number of accidents in a large number of factories, and proved that the number of accidents obeyed the statistical distribution of the extremely small probability of occurrence, and the probability of accidents for each individual was different. He calculated the correlation coefficient of the number of accidents in the first five months and the next five months in some factories, and the result was (0.04 ± 0.09)-(0.71 ± 0.06).

Maybe tracked and investigated a factory with 3 000 workers, and found that workers with no accidents happened in the first year had an average of 0.30 to 0.60 accidents in the following years; workers with an accident happened in the first year had an average of 0.86 to 1.17 accidents in the future; workers with two accidents happened in the first year had an average of 1.04 to 1.42 accidents in the future. All this fully proves accident proneness.

In 1939, Farmer and Chamber clearly put forward Accident Proneness. They believed that accident-prone tendencies caused mainly industrial accidents.

To determine if someone is accident-prone tendency, a series of psychological tests must be used. For example, the YG test (Yatabe-Guilford test) was used in Japan to test the personality of workers. In addition, Accident-prone tendency can also be found through observation of daily worker behavior. Generally speaking, accident-prone tendency tend to lose their minds when performing production operations, and cannot always focus on operations, so they cannot rapidly adapt to changing external conditions.

The theory believes that this group of people can be distinguished by their personality characteristics, and refuse to hire them to reduce accidents in industrial production. This theory attributed the cause of the accident to human nature. Some people still agree with this theory, but many subsequent research results have not confirmed the correctness of the theory.

5.2.2 Accident liability

After the Second World War, people put forward new insights of Accident Proneness. Some studies have shown that the notion that most industrial accidents are caused by accident-prone persons is wrong. Some people are more likely to have accidents than others, and it is related to the higher risk of their operations.

More and more people believe that the responsibility of accidents should not be simply regarded as workers' inattention, and attention should be paid to the important position of mechanical and material dangerous nature in the cause of accidents. As a result, the theory of Accident Liability emerged. Accident Liability refers to the tendency of certain personnel to be prone to accidents under certain production conditions.

A. Mint and Bloom (M. L. B) suggested replacing Accident Proneness with Accident Liability. They believed that accidents were not only related to personal factors, but also related to production conditions.

Based on this insight, W. A. Kerr investigated the relationship between 40 personal factors and production operating conditions in 53 electronic factories and the frequency of accidents and severity of injuries, and found the main factors affecting the frequency of accidents: short handling distances, severe noise, many temporary workers, poor worker consciousness, etc. The main factor related to the severity of the accident is the worker's "masculine" style, followed by lack of consciousness, lack of guidance, many elderly workers, discontinuous attendance and so on, which prove that the accident situation is closely related to the production operation conditions.

In fact, many operations in industrial production have certain requirements for the quality of the operators, or that the personnel have certain professional suitability. When the quality of the personnel does not meet the requirements of the production operation, errors or unsafe behaviors will occur in the production operation, resulting in accidents.

For example, in special operations, operators must undergo special training, strict assessment, and obtain qualifications for special operations. Therefore, although accident proneness theory is wrong to attribute the cause of industrial accidentsto a small number accident-prone tendencies, however, from the perspective of occupational suitability, there are certain merits regarding the accident proneness.

5.3 Accident causal chain theory

5.3.1 Causal inheritance

There is a certain causal relationship between the occurrence of accidents and their causes. "Cause" and "effect" are inherited, and the result of the previous paragraph is often the cause of the next paragraph.

The accident phenomenon is a "consequence" and has a certain relationship with its "antecedent". Causality occurs one after another at multiple levels; the first cause is the result of the second cause, the second cause is the result of the third cause, and so on.

Generally speaking, accident causes are often divided into direct causes and indirect causes. As shown in Figure 5-2, The direct cause, also called the primary cause, is the closest cause of the accident in time. Direct causes are usually further divided into two categories, physical causes and human causes. The cause of material is the unsafe state of equipment, materials, environment (also known as environmental material), etc.; the cause of human is the unsafe behavior of people. The indirect cause is the secondary, tertiary or even multi-level basic cause from the origin of the accident.

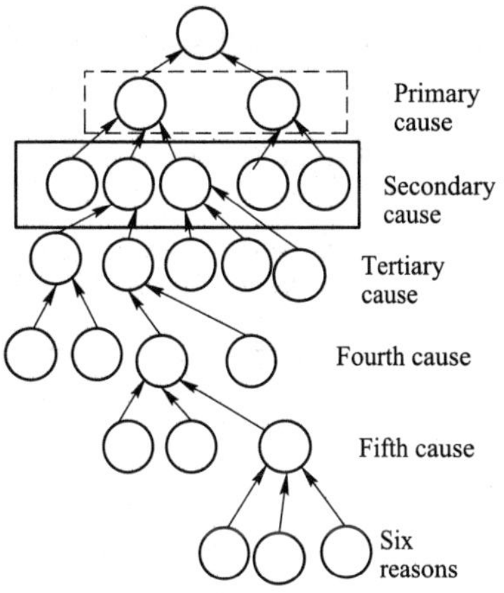

Figure 5-2 Causal inheritance

Indirect causes are roughly divided into 6 categories:

(1) Technical reasons: technical aspects such as the design, installation, and maintenance of the main mechanical equipment are not perfect, and there are technical defects in the process and protective equipment;

(2) Reasons for education: insufficient safety knowledge education for employees, insufficient training, lack of safety awareness among employees, etc.;

(3) Physical causes: refers to the operator's physical defects, such as vision or hearing impairment, and lack of sleep;

(4) Mental causes: refers to mental states such as restlessness, tension, fear, absent-mindedness, and psychological disorders or mental retardation;

(5) Reasons for management: corporate leaders are not strong in their sense of responsibility for safety, regulations and standards and inspection systems are not perfect, decision-making errors, etc.;

(6) Social and historical reasons: involving systems, policies, institutional relationships, local protectionism, institutions, systems, and the historical process of industrial development.

Among the indirect causes in items (1) to (6), (1) to (4) are secondary causes, and (5) to (6) are basic causes. The principle of causal inheritance can be regarded as the following chain "event chain": loss← accident←primary cause (direct cause)←secondary cause (indirect cause)←basic cause.

When tracing the cause of an accident, start with a retrograde investigation of the cause. Causality is inherited and is a multi-level chain relationship. The primary cause is the result of the secondary cause, and the secondary cause is the result of the tertiary cause, which can be traced back to the most basic cause.

If appropriate countermeasures are adopted and any one of the reasons is removed, this "chain of events" will be cut off and accidents can be prevented. But even if the direct cause is removed, as long as the indirect cause still exists, it will not prevent the occurrence of a new direct cause. Therefore, as the most fundamental countermeasure, it should be traced back to the secondary cause as well as the basic cause, and in-depth study to solve it.

5.3.2 Types of accident causal

Chain type: incident 1→incident 2→incident 3→incident 4→ an accident;

Causality: different incidents→ an accident;

Compound type (Figure 5-3).

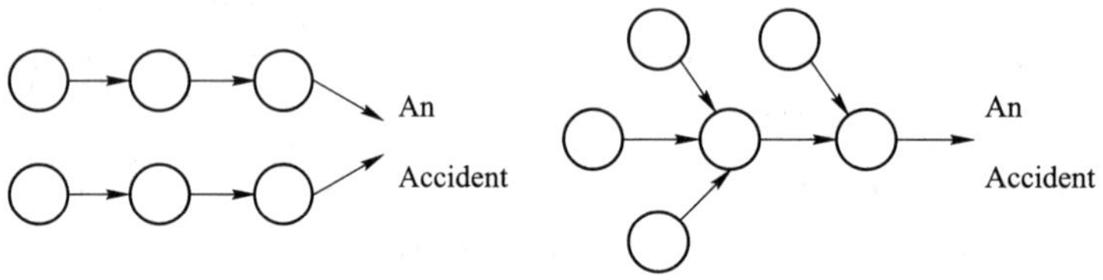

Figure 5-3 Compound type of accident causal

5.3.3 The causal and harmful object

The so-called causal object refers to the machinery, device, natural or man-made object, environmental object, etc. that caused the origin of the accident; the harmful object refers to the substance that directly caused the accident and harmed people. The unsafe state leads to the effect of the causal object; the injurious object is caused by the causal object to cause the consequences of the accident.

In terms of the series of things, from the distant cause to the near cause, from the earliest cause (cause 0) to the harming stuff (stuff 1), the stuff 1 will derive a new harming stuff (stuff 2), which continuously produces up to the accident phenomenon of personal injury caused by contact with people. The relationship between the causal and harmful object is shown in Figure 5-4.

For example:

During the welding operation, sparks splashed, ignited the polyurethane rubber, and the combustion products poisoned carbon monoxide; sparks splashed on the varnish gasoline and caused a fire and burned the workers. At the same time, the fire caused the gasoline barrel to explode, the barrel piece flew out and injured people.

The cause of this accident was the electric welding device, the harmful object 1 was sparks; the harmful object 2 was polyurethane rubber and gasoline; the harmful object 3 was CO, high-

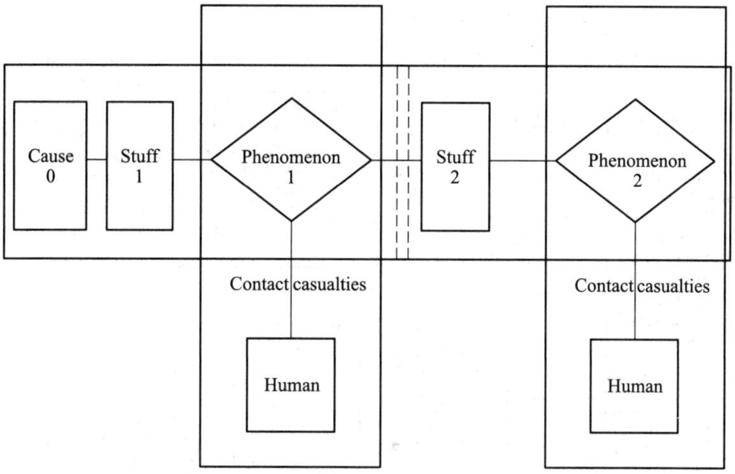

Figure 5-4 The relationship between the causal and harmful object

temperature combustibles, and gasoline barrel fragments. The causal relationship of thiscase is shown in Figure 5-5. The causality diagram of CO poisoning accident in a gold mine and its causal sequence diagram are shown in Figure 5-6 and Figure 5-7.

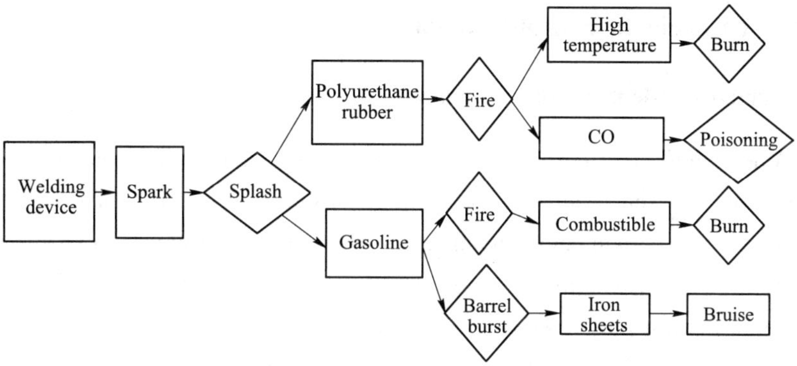

Figure 5-5 The relationship between the causal and harmful object in the welding operation

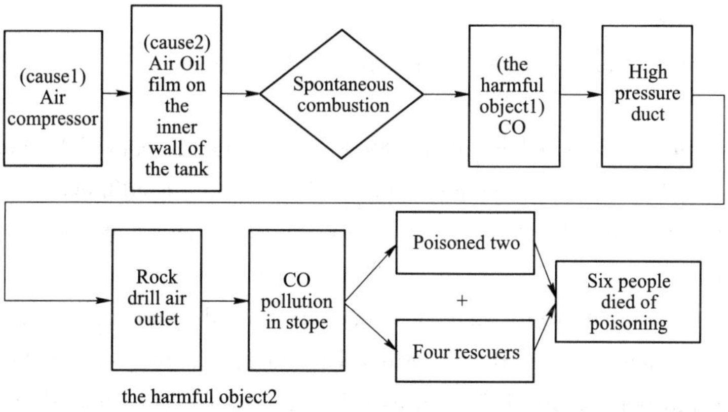

Figure 5-6 The causality diagram of the accident of CO poisoning in the 2nd well-270m9303 stope of a gold mine

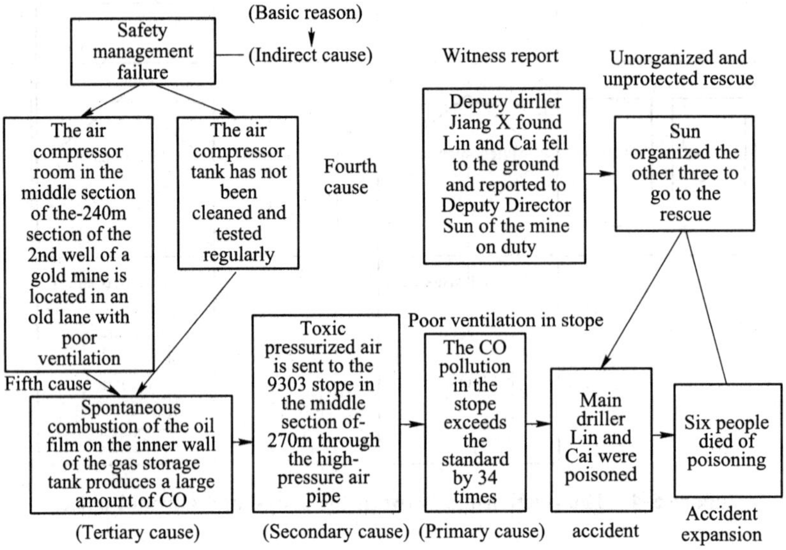

Figure 5-7 The causal sequence diagram of the accident of 6 deaths from CO poisoning in a gold mine stope

5. 3. 4 Heinrich accident causality theory

5. 3. 4. 1 Heinrich accident causality theory

In 1936, the American W. H. Heinrich proposed the accident causal chain theory.

Heinrich believes that the occurrence of an injury accident is a series of events that occur in sequence according to a certain causal relationship. He used five dominoes to vividly illustrate this causal relationship, that is, after the first one falls, it will cause a chain reaction of the following cards and fall, and the last one is an injury accident.

This theory is called the "domino" theory. The domino theory established the important concept of the chain of events that caused accidents, and provided a valuable method for latecomers to study the mechanism of accidents.

Heinrich has investigated 75 000 industrial accidents and found that 98% of them are preventable.

Among preventable work-related accidents, 89. 8% were mainly caused by unsafe behavior of people, and only 10. 2% were mainly caused by unsafe conditions of equipment and materials. According to this statistical result, most industrial accidents are caused by unsafe behavior of workers.

Heinrich also believes that even if some accidents are caused by the unsafe state of things, their unsafe state is also caused by workers' errors. Therefore, this theory, like the accident-prone personality theory, attributed most of the causes in the chain of events to workers' errors, showing the limitations of the times.

5.3.4.2 Domino Stock Theory

Domino Stock Chain Theory Model is displayed in Figure 5-8.

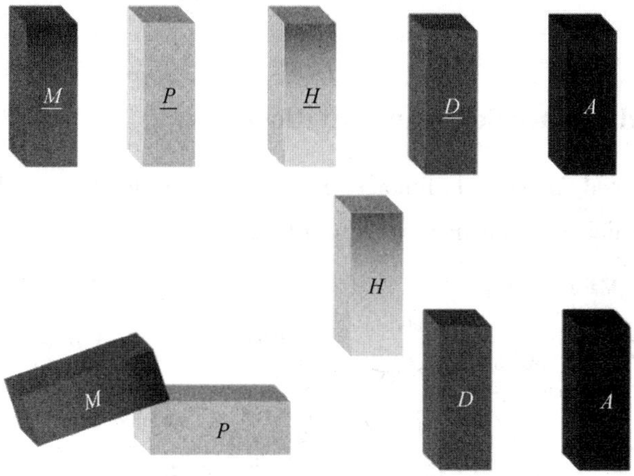

Figure 5-8 Domino Stock Chain Theory Model

The Mathematical Model of Domino Stock Theory:

If A_0 is used to represent the casualty accident, according to the domino stock model, the casualty accident A must meet the conditions that A, M, P, H, and D all fall, that is, the event represented by these five dominoes All happens (and gate), so:

$$A_0 = A \cdot M \cdot P \cdot H \cdot D \tag{5-1}$$

From this, we can get:

$$P[A_0] = P[A] \cdot P[M] \cdot P[P] \cdot P[H] \cdot P[D] \tag{5-2}$$

The probabilities of the five events A, M, P, H, and D are all <1, so $P[A_0] << 1$, indicating that the probability of casualties is very small. If one of the cards is drawn (such as $H=0$), there are:

$$P[A_0] = P[M] \cdot P[P] \cdot 0 \cdot P[D] \cdot P[A] = 0 \tag{5-3}$$

So A_0 is an impossible event, and casualties will not happen.

5.3.4.3 Other accident causal chain theories

The accident chain model similar to Heinrich's causal chain theory includes:

(1) Bode accident causality theory;

(2) Adams accident causality theory;

(3) The Causal Chain Theory of Toruzo Kitagawa.

5.3.5 Bode accident causality theory

On the basis of Heinrich's accident causal chain theory, Bode proposed an accident causal chain theory that is more consistent with modern safety viewpoints.

Bode's accident causal chain process is also five factors, but the meaning of each factor is different from Heinrich's:

(1) Management defect;

(2) Personal reasons;

(3) Direct reasons;

(4) An Accident;

(5) Loss.

5.3.6 Edward Adams accident causality theory

Edward Adams proposed a causal linkage model similar to the Bode accident causality theory (Figure 5-9), which is given in tabular form (Table 5-1).

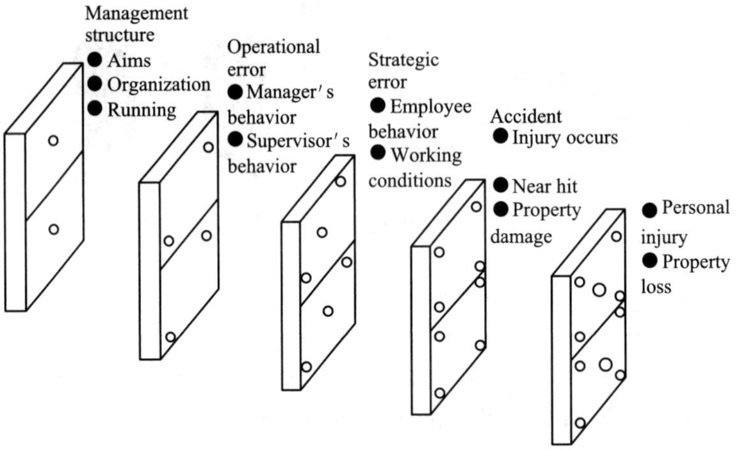

Figure 5-9 Edward Adams accident causality theory

Table 5-1 Edward Adams accident causality theory

Management system	Management mistake		Site error	An accident	Injury or damage
Aims; Organization; Function	Problems of the manager: Policy; Aims; Specification; Responsibility; Rank; Assessment; Permission granted	Mistakes of technical staff: Behavior; Responsibility; Scope of authority; Rules; Guide; Initiative; Positive; Business Activity	Unsafe behavior; Unsafe state	Casualties; Damage accident; No injury accident	To people; To object

5.3.7 Toruzo Kitagawa causal chain theory

The previous several accident causal chain theories limited the scope of investigation within the enterprise. In fact, the causes of industrial injury accidents are very complicated. Many social factors such as politics, economy, culture, education, and technological level of a country or region have an important influence on the occurrence and prevention of injury accidents.

Based on this consideration, Japan's Toruzo Kitagawa made certain amendments to Heinrich's theory and proposed another accident causal chain theory, which is shown in Table 5-2.

Table 5-2 Toruzo Kitagawa causal chain theory

Basic reason	Indirect cause	Direct cause		
School education reasons; Social Causes; Historical reasons	Technical reasons; Educational reasons; Health reasons; Mental reasons; Management reasons	Unsafe behavior; Unsafe state	Accident	Harm

5.3.8 The Theory of causes of accidents after World War Ⅱ

After the outbreak of the World War Ⅱ, the emergence of new military equipment such as high-speed aircraft, radars, and automatic artillery has brought complexity and tension to operations, making it difficult for people to adapt, and mistakes often occur. As a result, a discipline specializing in the study of human working ability and its limitations-ergonomics has emerged.

Ergonomics has had a profound impact on the development of industrial safety after the war. Its rise marked a major change in the relationship between man and machine in industrial production. In the past, workers were trained according to the characteristics of machinery to allow workers to meet the requirements of machinery. Now, machinery is designed according to the characteristics of people and machinery to make machinery suitable for human operations.

In the human-machine relationship, the concept of putting people first and making machines fit people has prompted people to re-understand the causes of accidents.

More and more people believe that the occurrence of accidents should not be simply described as worker's character defects or carelessness. We should pay attention to the role of mechanical and material hazards in accidents, and emphasize the realization of production conditions and the inherent characteristics of mechanical equipment. Only by safety can the occurrence of accidents be effectively and effectively reduced, and theories about the cause of accidents such as "epidemiological methods for accidents" and "abnormal transfer of energy" have emerged.

5.4 Epidemiological methods used in accidents

In 1949, Gorden used the mechanism of epidemic infection to describe the mechanism of accident and proposed epidemiological methods used in accidents.

There are three main types of epidemiological analysis:

(1) The characteristics of the person (patient), such as age, gender, psychological status, immunity, etc.

(2) Environmental characteristics, such as temperature, humidity, season, community sanitation, epidemic prevention measures, etc.

(3) The characteristics of pathogenic vectors, such as viruses, bacteria, mycoplasma, etc. The

interaction of these three factors can lead to human diseases.

Similarly, for accidents, one must consider human factors, second, consider operating environment factors, and third, consider the medium that caused the accident.

This epidemiological method considers the age, gender, physical and psychological conditions of the person (accident victim), and environmental characteristics, such as work and living areas, social conditions, and seasons, as well as the characteristics of the media, such as in epidemiology viruses and bacteria.

However, in industrial accidents, it is no longer a biological problem with a certain scope. Instead, the "media" should be understood as the energy that contributes to the accident, that is, the source of injury, such as mechanical energy, potential energy, electrical energy, thermal energy, and radiation energy.

Energy, like viruses, is a transient cause of accident or disease phenomena. However, the vector of disease is always absolutely harmful, but the degree of harmfulness varies.

Energy is a favorable motive force most of the time, and it is a function that serves production. Only when the energy flows back to the human body by accident, it is the origin and medium of the accident.

The epidemiological method is a great improvement over the early accident theory that only considers human errors. It clearly puts forward the characteristics of the relationship between causal factors.

The theory recognizes that accidents are the result of the interaction of certain factors among the three sets of variables (the characteristics of the parties, the environmental characteristics, and the energy characteristics of the medium).

The disadvantage of this theory is that the three groups of variables contain a lot of content that needs to be studied. These many factors must be counted and evaluated by a large number of specimens, but they lack clear guidance.

5.5 Accidental energy release theory

5.5.1 Overview

It was proposed by Gibson in 1961 and added by Hadden in 1966.

The theory of "abnormal transfer of energy" is an important step in the development of the theory of accident causes.

The theory believes that accidents are an abnormal or undesirable energy transfer, and various forms of energy constitute the direct cause of injury accidents. Therefore, energy or energy carriers should be controlled to prevent injury accidents. The effective measure to prevent abnormal energy transfer is to shield the energy.

The development of modern industry originated from the conversion of chemical energy of fuel into thermal energy, and the conversion of water as a medium into steam. Then the thermal energy of steam was converted into mechanical energy and transported to the production site. This is the energy conversion of the steam engine power system.

In the electrical age, the potential energy of water or the kinetic energy of steam is converted intoelectrical energy, and the electrical energy is converted into mechanical energy at the production site for product manufacturing and processing.

Nuclear power plants use atomic energy to convert into electrical energy.

In short, energy is a physical element with the ability to do work, and it is the most basic physical quantity that constitutes a system of matter and field.

When humans use energy, they must take measures to control energy, so that energy can be generated, converted, and performed according to people's intentions.

From the perspective of energy flow in the system, the energy should be controlled to flow in accordance with the energy flow channels prescribed by people.

If the control of energy is lost for some reason, an accidental release or escape of energy against people's wishes will occur, which will cause an accident to stop the activity in progress.

If the accidentally released energy acts on the human body, and the effect of the energy exceeds the human body's ability to withstand, it will cause personal injury. Table 5-3 shows the damage to the human body when it is subjected to various forms of energy that exceed its capacity. If the accidental released energy acts on equipment, buildings, objects, etc. , and the effect of the energy exceeds their resistance capacity, it will cause damage to equipment, buildings, and objects.

Table 5-3 Energy type and damage

Type of energy applied	Primary injury	Examples and notes
Mechanical energy	Displacement, tearing, rupture and crushing, mainly injuring tissue	Injuries caused by collisions with moving objects such as bullets, hypodermic needles, knives and falling objects, as well as injuries caused by collisions with relatively stationary equipment by a moving body, such as falling, flying and car accidents. The specific injury result depends on where and how the combined force is applied. Most injuries are of this type
Thermal energy	Inflammation, coagulation, scorching and incineration, hurting any level of the body	First-degree, second-degree and third-degree burns, the specific injury results depend on the location and method of heat action
Electricity	Interference with nerve-muscle function and coagulation, scorching and incineration, and damage any level of the body	Death by electrocution, burns, interference with nerve function, as in electroconvulsive therapy. The specific damage result depends on the part and method of electric energy action
Ionizing radiation energy	Destruction of cellular and subcellular components and functions	Reactor accidents, therapeutic and diagnostic exposures, abuse of the role of isotopes and radioactive elements. The specific damage result depends on the part and method of taking the radiant energy
Chemical energy	Injuries generally depend on the specific substances of each or each organization	Including injuries caused by animal andphytotoxins, chemical burns such as potassium hydroxide, bromine, fluorine and sulfuric acid, and most elements and compounds in sufficient doses that produce less severe and many types of injuries

The energy types such asmechanical energy, thermal energy, electricity, ionizing radiation energy, chemical energy can lead to abnormal transfers of energy and eventually accidents.

5.5.2 Principles and measures to prevent accidental release of energy

Starting from the theory of accidental release of energy, the prevention of injury accidents is to prevent the accidental release of energy or hazardous substances and prevent the human body from contacting excessive energy or hazardous substances.

Haddon believes that shielding protection systems can be used as safety measures to prevent energy transfer to the human body. He called the measures to restrain and limit energy and prevent the human body from contacting energy as shielding.

This is a broad shield. Under certain conditions, whether a certain form of energy can cause injury and cause casualties should depend on: (1) the magnitude of human contact energy; (2) contact time and frequency; (3) the concentration of energy; (4) sooner or later the barrier is set up, the earlier the barrier is set up, the better the effect.

According to the amount of energy, it is possible to study whether to establish a single shield or multiple shields (redundant shields).

Typical types of protection energy flowing back to the human body:

(1) Limit energy: limit the size and speed of energy, stipulate the safety limit, and try to use low-energy technology or equipment in the production process. Such as limiting the driving speed, stipulating low-voltage electricity for mine lighting, etc.

(2) Use safer energy to replace dangerous energy: sometimes the used energy is more dangerous, so consider using safer energy instead. Such as hydraulic coal mining instead of blasting, the application of carbon fire extinguishing agent instead of carbon tetrachloride and so on.

(3) Prevent energy accumulation: for example, control the concentration of explosive gas, and try not to empty the chute (reduce and release potential energy).

(4) Control energy release: establish protective devices to control accidental release of energy. Such as the use of protective containers (such as pressure-resistant oxygen tanks, special containers for holding radioactive isotopes) and the living area away from pollution sources.

(5) Delaying energy release: slowly releasing energy can reduce the energy released per unit time and reduce the effect of energy on the human body or facilities. Such as the use of safety valves, escape valves, vibration absorption devices, etc.

(6) Open up channels for releasing energy: safely release energy through new energy release channels. Such as grounding wires, exhausting gun smoke through local ventilation devices, etc.

(7) Set up shielding facilities: shielding facilities are physical entities that prevent people from contacting energy, that is, shielding in a narrow sense. Shielding facilities can be set up on energy sources, such as shock wave absorbing chambers, mufflers, and atomic shields, etc.. It can also be set on personnel, such as safety helmets, safety shoes, gloves, masks and other personal protective equipment.

(8) Set up a barrier between people, things and energy: isolate energy from people and things in time and space. Such as protective covers, fire doors, closed doors, waterproof gate walls, etc.

(9) Improve protection standards: such as the use of double insulated tools, continuous monitoring

and remote control.

(10) Change the process flow: change the unsafe process into a safe process. Such as replacing highly toxic substances with non-toxic and less toxic substances.

(11) Rehabilitation or first aid: treatment, correction, and reduction of the degree of injury or restoration of original functions. Limit the scope of disasters to prevent expansion of losses. Do a good job in first aid, conduct self-rescue education, etc.

(12) Shielding in the form of information: shielding in the form of information such as various warning measures can prevent unsafe behaviors of personnel or avoid human errors, and prevent personnel from contacting energy.

5.5.3 Accident causal chain model based on energy viewpoint

When investigating the causes of casualties, Michael Zabetakis of the U. S. Bureau of Mines found that most casualties were caused by excessive energy or accidental release of dangerous substances that interfere with the normal energy exchange between the human body and the outside world. And this excessive release of energy or hazardous substances is caused by unsafe behaviors of people or unsafe conditions of things. That is, the unsafe behavior of people or the unsafe state of things makes energy or dangerous substances out of control, which is the fuse for the release of energy or dangerous substances.

Based on the accidental release theory of energy, Zabetakis established a new accident causal chain model, as shown in Figure 5-10.

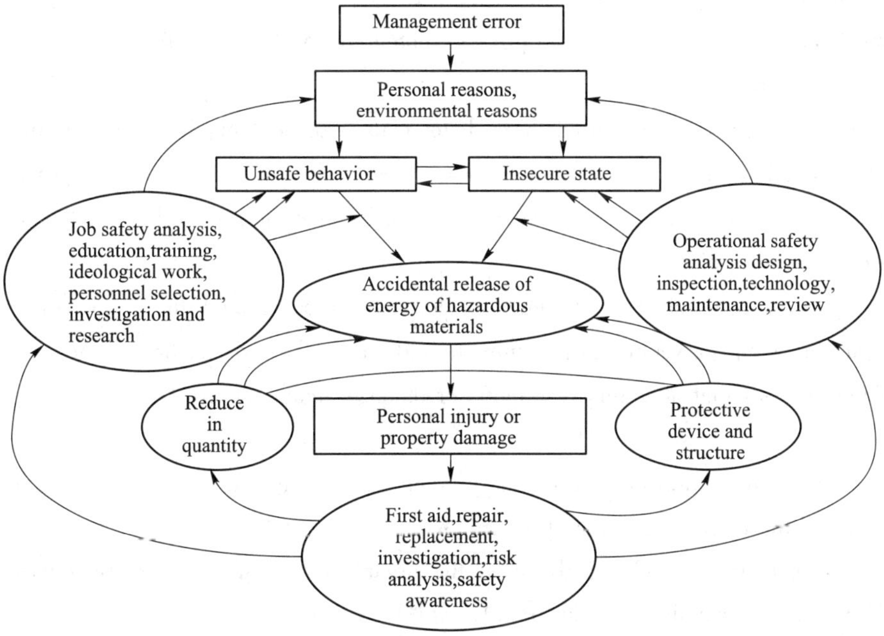

Figure 5-10 Accident causality chain model from the energy perspective

(1) Accident: an accident is the accidental release of energy or dangerous substances, which is the direct cause of injury. To prevent accidents, technical improvements can be used to prevent accidental release of energy, education and training can be used to improve the ability of employees to recognize dangers, and personal protective equipment can be worn to avoid injuries.

(2) Unsafe behavior and unsafe state: the unsafe behavior of people and the unsafe state of things are the direct causes of accidental release of energy. They are lack of management, ineffective control, lack of knowledge, errors in the estimation of existing risks, or Symptoms of basic causes such as other personal factors.

(3) Basic reasons: the basic reasons include three aspects.

1) Safety policies and decisions of business leaders. It involves production and safety goals; staff configuration; information utilization; responsibilities and terms of reference, staff selection, education and training, arrangement, guidance and supervision; information transmission, equipment, equipment and equipment procurement and maintenance; normal and abnormal times Operating procedures; equipment maintenance, etc.

2) Personal factors. Ability, knowledge, training; motivation, behavior; physical and mental state; reaction time, etc.

3) Environmental factors. Factors such as natural conditions and natural environment.

5.5.4 Two types of hazard theory based on energy viewpoint

The term " dangerous source" is Hazard in English, and the word means " Hazard-a source danger", which means the source of danger. Hazard is also translated into "hazard". W. Hammer (Willie Hammer) defines hazards as "potential unsafe factors that may cause personal injury or financial loss."

The hazard source is a potential hazard to cause anaccident, it is a certain potential environmental condition beyond the direct control of people. In a working environment where work-related injuries or occupational diseases occur, that is, the operation of workers in the workplace, there are many sources of danger.

In system safety research, the existence of hazards is considered to be the root cause of accidents, and accident prevention is to eliminate and control hazards in the system.

McFarland believes that all injury accidents (or damage accidents) are due to the following two energy factors:

(1) Exposure to some form of excess energy (such as electricity exceeding 220V) that exceeds the resistance of the body's tissue (or structure);

(2) The normal energy exchange between the organism and the surrounding environment is disturbed (such as suffocation, drowning, etc.).

According to this viewpoint, the damage caused by energy can be divided into the first type of injury and the second type of injury.

5.5.4.1 First category hazard

According to the theory of accidental release of energy, an accident is an accidental release of energy or dangerous substances. Excessive energy acting on the human body or dangerous substances that interfere with the exchange of energy between the human body and the outside world is the direct cause of personal injury. Therefore, the energy or hazardous substances that may be accidentally released in the system are called the first type of hazard source.

Each part of the human body has a certain resistance to the action of each energy, that is, there is a certain injury threshold. When a certain part of the human body is in contact with a certain energy, whether it can be injured and the severity of the injury depends mainly on the energy level of the human body.

The more the energy acting on the human body exceeds the damage threshold, the greater the possibility of causing damage. For example, when a spherical projectile hits the human body with an impact force of 4.9 N, it will at most slightly scratch the skin, while when a spherical projectile strikes a human head with an impact force of 68.9 N, it will cause a human skull fracture.

Common first category hazards are:

(1) Devices and equipment that generate and supply energy;

(2) Devices, equipment, and places that enable human bodies or objects to have high potential energy;

(3) Energy carrier;

(4) Devices, equipment, and places that may generate huge energy once they are out of control, such as chemical plants with a strong exothermic reaction;

(5) Devices, equipment, and places where energy accumulation or sudden release may occur once out of control, such as various pressure vessels;

(6) Dangerous substances, such as various toxic, harmful, combustible and explosive substances, etc.;

(7) Installations, equipment and places for the production, processing and storage of hazardous substances;

(8) Objects that will cause accidental release of human energy once the human body comes into contact with it.

5.5.4.2 The second hazard

In production and life, in order to use energy and let it flow, convert and perform work in the system according to people's intentions, measures must be taken to restrain and limit energy, that is, dangerous sources must be controlled. The shield that restrains and limits energy should reliably control energy to prevent accidental release of energy.

In fact, absolutely reliable control measures do not exist. Under the complex action of many factors, the control measures to restrain and limit energy may fail, and the energy shield may be destroyed and accidents may occur.

Various unsafe factors that lead to the failure or destruction of restraint and energy limiting measures are called the second category of hazards, including three aspects: people, objects, and the environment.

(1) Human error: it means that the result of human behavior deviates from a predetermined standard, and human unsafe behavior can be regarded as a special case of human error. Human error may directly destroy the control of the first category of hazard sources and cause accidental release of energy or hazardous materials.

(2) Obstacle of objects: refers to the phenomenon that the predetermined functions cannot be realized due to low performance, and the unsafe state of objects can also be regarded as a fault state. The failure of the object may directly invalidate the measures to restrain, limit energy or hazardous substances and cause an accident. For example, electrical leakage occurs due to damage to the insulation of the wire; the ruptured pipeline causes the leakage of toxic and harmful media, and so on.

(3) Environmental factors: mainly refers to the environment in which the system operates, including physical environments such as temperature, humidity, lighting, dust, ventilation, noise, and vibration, as well as the soft environment of the enterprise and society. A poor physical environment can cause physical failure or human error. For example, a humid environment will accelerate corrosion and reduce the strength of the cotton structure or container; strong noise in the workplace affects people's emotions, distracts people's attention and causes human errors.

5.5.4.3 The theory of causes of two types of hazardous source accidents

The occurrence of an accident is the result of two types of hazards working together. The existence of the first type of hazard is the prerequisite for the occurrence of an accident. Without the first type of hazard, there is no accidental release of energy or hazardous materials, and there is no accident. In addition, if there is no second-class hazard source that destroys the control of the first-class hazard source, accidental release of energy or hazardous materials will not occur. The appearance of the second type of hazard is a necessary condition for the first type of hazard to cause an accident. The accident causality chain of two kinds of hazard theory is shown in Figure 5-11. And Figure 5-12 demonstrates the hierarchical network accident cause model.

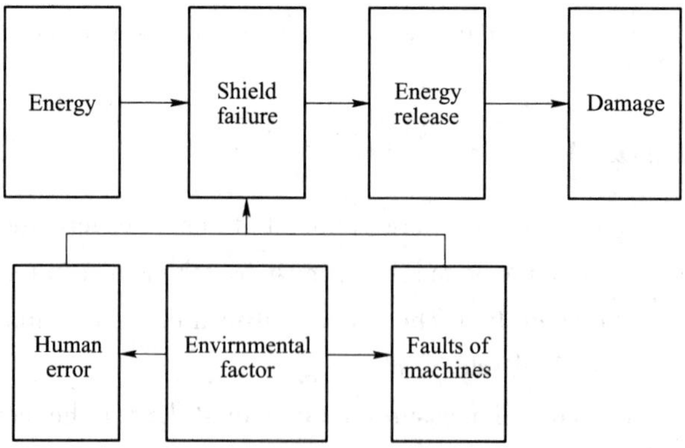

Figure 5-11 The accident causality chain of two kinds of hazard theory

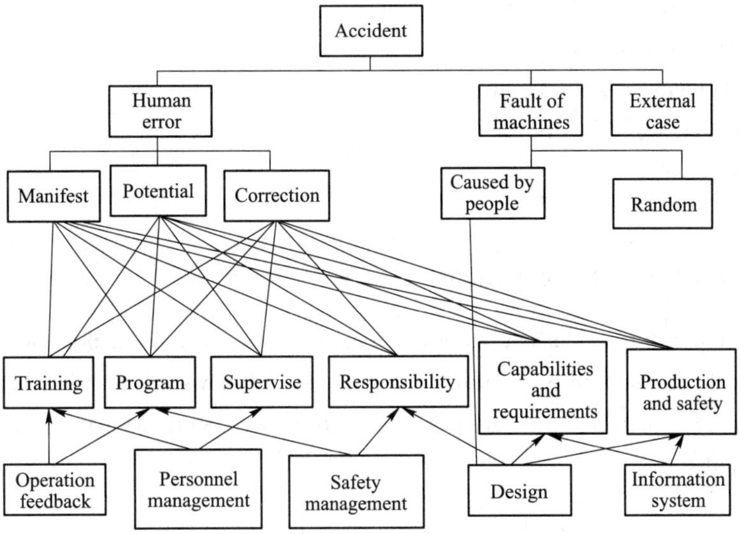

Figure 5-12 Hierarchical network accident cause model

During the occurrence and development of an accident, the two types of hazards are interdependent and complementary: the energy released by the first type of hazard during an accident is the main body of energy that causes personal injury or property damage, and determines the severity of the consequences of the accident; The difficulty of occurrence of such hazards determines the possibility of accidents. The two types of hazards jointly determine the damage of the hazard.

5.6 The main cause of human error

In the past 30 years, science and technology have continuously developed and progressed, and production processes and production equipment have become more and more complex. Information theory, system theory, and cybernetics have matured one after another, and have been widely used in various fields.

For the safety of complex systems, the previous theories and methods cannot be solved well. Therefore, new safety theories and methods have emerged.

The main cause theory of human error has a basic point of view in this type of accident theory, that is, human error will lead to accident, and human error is caused by the error of human response to external stimuli (information).

These theories all start from the point of view of whether human characteristics are matched and coordinated with machine performance and environmental conditions. They believe that machine and environmental information is constantly reflected to the brain through human senses. If people can correctly recognize, understand, and judgeby making correct decisions and taking appropriate measures, danger can be turned into a breeze and accidents, injuries and deaths can be avoided.

Conversely, if people fail to perceive and recognize the danger they are facing, or if their judgment is not accurate and fail to adopt the correct way of behavior, accidents and casualties will occur.

The more famous human error accident models:

(1) Wigglesworch accident model;

(2) J. Surry accident model diagram;

(3) Lawrence Model;

(4) Anderson model;

(5) Haier model.

5.6.1 Wigglesworch accident model

There are many types of accident causes. Wigglesworch proposed in 1972 that there is an accident cause that forms the basis of all types of injuries. Figure 5-13 shows this accident model. This cause is "human error". He defined "human error" as "a person responds wrongly or inappropriately to an external stimulus".

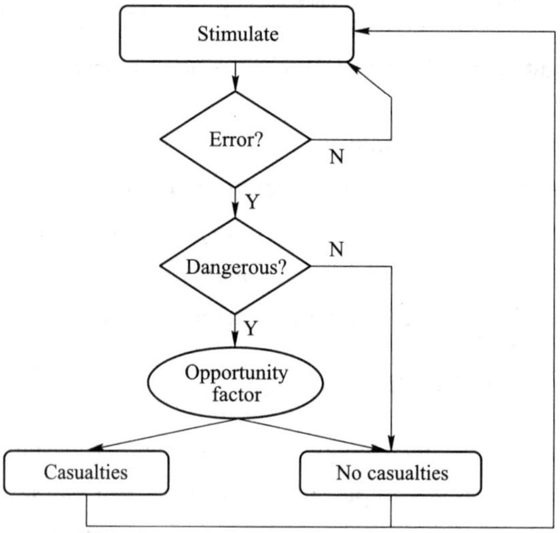

Figure 5-13 Wigglesworch's accident model based on human error

During the production and operation of workers, various "stimuli" continue to appear. If workers respond correctly or appropriately, accidents will not occur.

That is, if there is no danger, there will be no accidents that accompany injuries; on the contrary, if there are incidents of human error, there is a possibility of accidents. Whether an accident can cause harm depends on various random factors. It may cause casualties or accidents without casualties.

5.6.2 J. Surry accident model diagram

An accident model that describes the causality of accidents based on the process of human information processing proposed by J. Surry in 1969 and developed in the early 1970s. This theory believes that people make mistakes in the process of information processing, leading to human behavior errors and accidents. The surry accident model is shown in Figure 5-14.

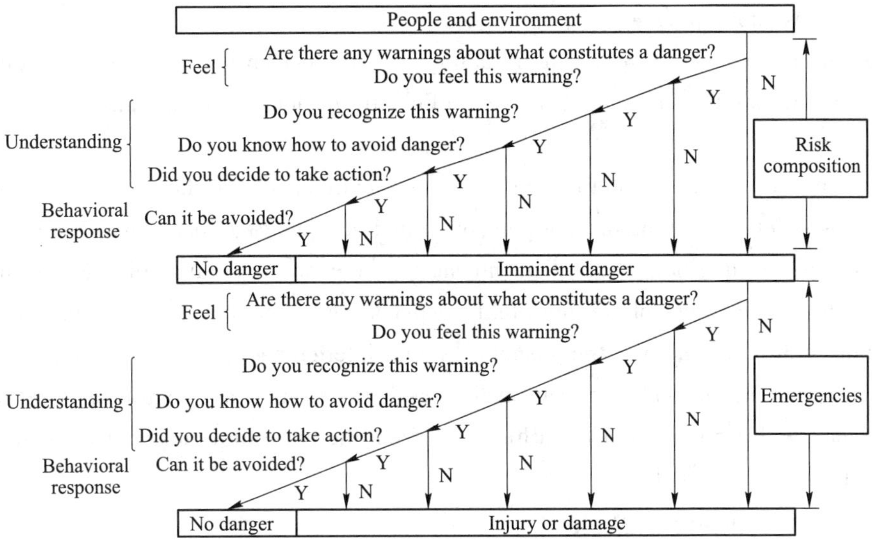

Figure 5-14 J. Surry accident model

5.6.2.1 Two stages

J. Surry divides the accident process into two stages: danger emergence and danger release. Each of these two stages includes a set of similar human information processing processes, namely, perception, cognition and behavior response processes.

At the emergence stage of danger, if every link of human information processing is correct, the danger can be eliminated or controlled; conversely, as long as any link has problems, the operator will be directly exposed to danger.

In the danger release stage, if all aspects of the human information processing process are correct, although the danger has already appeared, the release of the danger can still be avoided without causing harm or damage. On the contrary, as long as any link is required. If something goes wrong, the danger will turn into injury or damage.

5.6.2.2 The information processing process of the J. Surry model

A Are there any warnings about the occurrence (or exposure to danger) of danger

The warning here means whether there is a noticeable difference between safe operating conditions in the work environment.

If the danger does not bring about a perceivable difference, it will expose people directly to the danger.

In actual production, even if the danger exists, it does not necessarily appear directly.

The enlightenment of this question is to let the unobvious dangerous state be fully revealed, which often requires certain technical means and methods to achieve this.

B Feel the warning

This question has two implications:

(1) The first is how the person's sensory ability is. If the person's sensory ability is poor or the attention is elsewhere, then even if there are sufficiently obvious warning signs, they may not be noticed;

(2) The second is the "interference" of the environment to the warning signal. If the interference is severe, it may hinder the perception and acceptance of dangerous information.

According to this question, the enlightenment is: there are individual differences in sensory ability, and the improvement of sensory ability depends on experience and training. At the same time, training can also improve the operator's anti-interference ability. In severe interference situations, use warning methods that can avoid interference (such as the use of light signals or acoustic signals with large differences in frequency from the noise in a noisy place) or increase the strength of the warning signal.

C Are you aware of this warning

This question asks whether the operator understands the meaning of the warning after feeling the warning.

The operator actually compares the warning information with the knowledge already in his mind to identify the existence of danger.

D Know how to avoid danger

The question is whether the operator has the knowledge and skills to avoid dangerous behavioral responses.

In order to make this knowledge and skills perfect and systematic, which is more conducive to taking correct actions, operators should receive corresponding training.

E Decided to take action

On the surface, there is no doubt about this issue. Since it is dangerous, of course we must take action. However, in actual situations, people's actions are driven by the dominant motive among various motives. The "risk-averse" motives for taking actions to avoid risks often intertwine with the "profit-seeking" motives (such as saving time, effort, making more money, enjoyment, etc.).

When the profit-seeking motive becomes the dominant motive, even though the operator recognizes the existence of danger and knows how to avoid it, the operator still "haves a fluke" and does not take risk-averse actions.

F Can the danger be avoided

The question is whether the operator can react quickly, swiftly and correctly after making the decision to take action.

Among the above six questions, the first two questions are related to people's perception of information, the third to fifth questions are related to people's cognition, and the last question is related to people's behavioral responses.

These 6 questions cover the whole process of human information processing, and reflect that there are many opportunities for mistakes and accidents in this process.

5.6.2.3 The reaction time problem of the J. Surry model

The J. Surly model is mainly suitable for describing situations where dangerous situations appear slowly, and accidents may occur if they are not corrected in time.

The reaction time of a normal person is 900 milliseconds. Therefore, a reaction time of 1 second or less allows a person to avoid danger in most cases. However, the reaction time of a person sometimes exceeds the critical time, and then danger cannot be avoided. Therefore, it also has certain reference value for describing rapid accidents.

5.6.3 Lawrence model

Lawrence based on the human error model of Wigglesworch, J. Surry and others, through the study of accidents in South African gold mines, in 1974, he proposed the main cause of human error for gold mining enterprises (Figure 5-15). This model has good practical value for more complicated accidents in general mining enterprises and other enterprises.

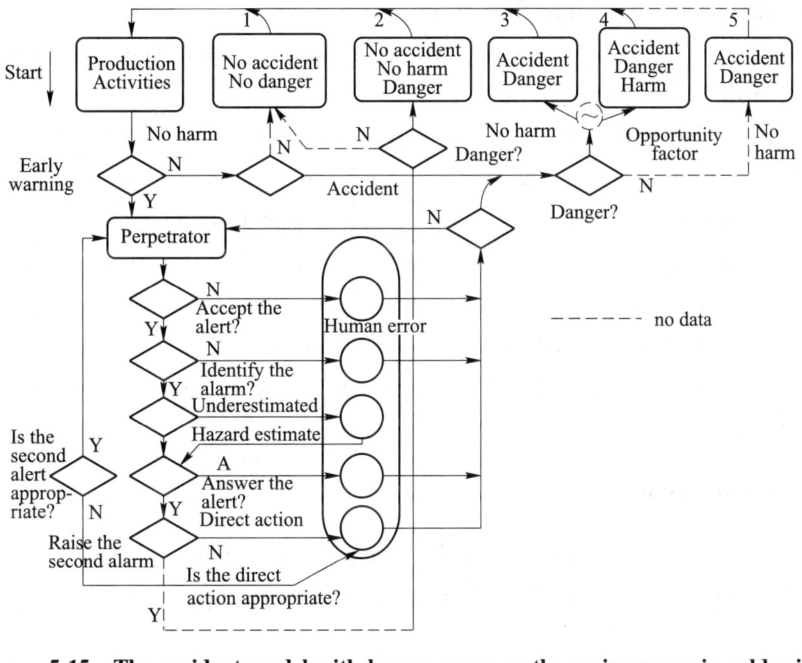

Figure 5-15 The accident model with human error as the main reason in gold mine

In the mining industry, continuous production activities including human factors may cause two results: injury and non-injury. Therefore, the definition of "accident" is: an unexpected event that interrupts normal production activities. The term accident is often used as a synonym for injury in mine safety work. However, whether an accident occurs or not depends on the dangerous situation (the probability of human injury) and chance factors. Table 5-4 lists eight theoretical combinations of accidents, dangers and injuries.

Under normal production conditions, there are no dangerous signs and unsafe information, that is, there is no initial warning. If there is no accident, there will be no interruption of production, and the result is "no accident, no danger, no harm", which belongs to type 1.

Table 5-4 Acombination of accident, danger and injury

Type	Accident	Danger	Injury
1	NO	NO	NO
2	NO	YES	NO
3	YES	YES	NO
4	YES	YES	YES
5	YES	NO	NO
Impossible to appear	YES	NO	YES
Impossible to appear	NO	YES	YES
Impossible to appear	NO	NO	YES

The horizontal line of "initial warning" is to the right, and accidents occur without the initial warning. This will produce 3, 4, and 5 types of results according to the occurrence of the danger and the chance factors of the injury. If there is danger, it will produce type 3 and 4 results; if there is no danger, it will produce type 5 results.

During the operation of the miner, roof pressure activities, water inrush signs, toxic gas gushing, visual and auditory sound and light signals; or from relevant instructions, regulations, standards, and mining processes that are compatible with safe production and environmental conditions Various "stimuli" of written information continue to appear.

If the worker's response is correct or appropriate, there is no danger and accidents accompanied by injury will not occur. On the contrary, if the worker responds improperly to the stimulus, there will be incidents of human error. This kind of chance factor may cause casualties or accidents without casualties.

5.6.4 Anderson model

Actually, the Surly model studies the interrelationship, feedback, and adjustment control between humans and hazards when potential hazards exist objectively (in the operation and environment of machinery).

However, the Thurley model did not explore why potential hazards are generated, and did not involve the operation of the machine and its surrounding environment.

In 1978, Anderson and others applied the Thurley model to analyze 60 industrial accidents and discovered the above-mentioned problems, which expanded the Thurley model and formed the Anderson model at last (Figure 5-16). This model adds a set of questions on the Surly model, which involves: the source and perceptibility of hazard clues, fluctuations in the operating system (instability of the mechanical operation process and environmental conditions), and control or Reduce these fluctuations to make them consistent with the fluctuations in human (operator) behavior. This increase in the work process makes the Seri model more useful.

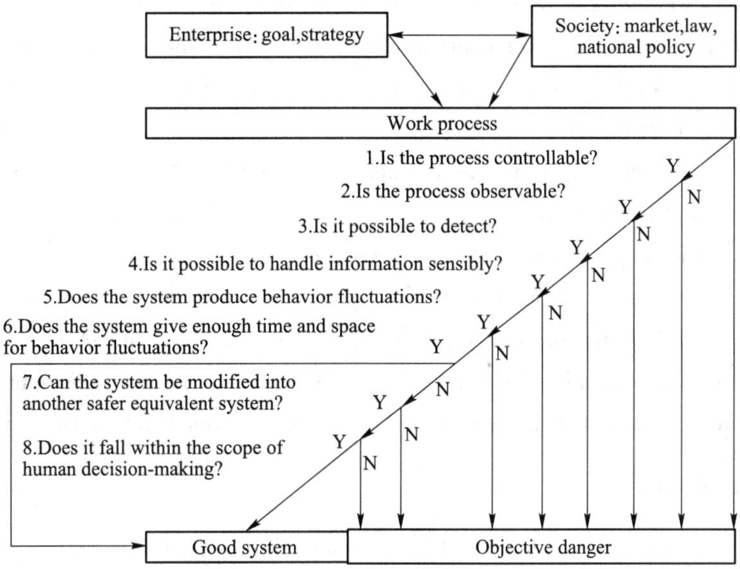

Figure 5-16 Anderson accident model

There are eight questions raised by the Anderson model:

(1) Is the process controllable?

That is to say, the danger caused by uncontrollable processes (such as lightning) cannot be avoided. This model discusses work processes that can be controlled.

(2) Is the process observable?

It refers to the ability to observe and understand the work process by relying on human senses or by means of instrumentation equipment.

(3) Is detection possible?

It refers to whether noise, poor lighting, fences, etc. in the work environment will hinder the observation and understanding of the work process.

(4) Is it possible to deal with information rationally?

This question has two meanings: one is to ask the operator whether he knows how the system works, if the system is not working properly, can he feel and recognize this situation. The other is to ask the fatigue caused by the system operation, mental stress (such as being in a state of high mental stress for a long time) and weakened attention will hinder their accurate observation and understanding of the system's working status.

The meaning of the above question is similar to the meaning of the first stage of the Surly model. The difference is that the Anderson model is for the entire system, while the Surly model is only for specific hazard clues.

(5) Does the system produce behavior fluctuations?

What is the instability of the operator's behavioral response? Is there any instability? How old is it?

(6) Does the operating system provide enough time and space for behavior fluctuations?

The question is whether the operating system (machinery, environment) has enough time and space to adapt to the instability of the operator's behavior. If it is, then it can be considered that the operating system is safe (the problem 7 and 8 are crossed in the figure, and the system is good), otherwise, go to the next question.

(7) Whether the system can be modified (machine or program)?

The aim is to adapt to the instability of the operator's behavior within the expected range.

(8) Is it within the scope of human decision-making?

Refers to whether the modification of the system can be decided by the operation and management personnel. Although the system can be changed to a safe one, it is impossible to modify the system if the operation and management personnel do not have the right to change, or if it involves policies and laws that are not within the scope of human decision-making.

5.6.5 Haier model

In 1970, Haier believed that accidents will occur when people fail to respond appropriately to the true circumstances of the incident, but they do not necessarily cause harm. Haier's model focuses on the interaction between the operator and the operating system. See Figure 5-17 for details.

The model is a closed-loop feedback system that clearly shows the interrelationships of the following four aspects: (1) Perceive the situation and receive information; (2) Process information; (3) Change the situation with actions; (4) New perception, processing, response.

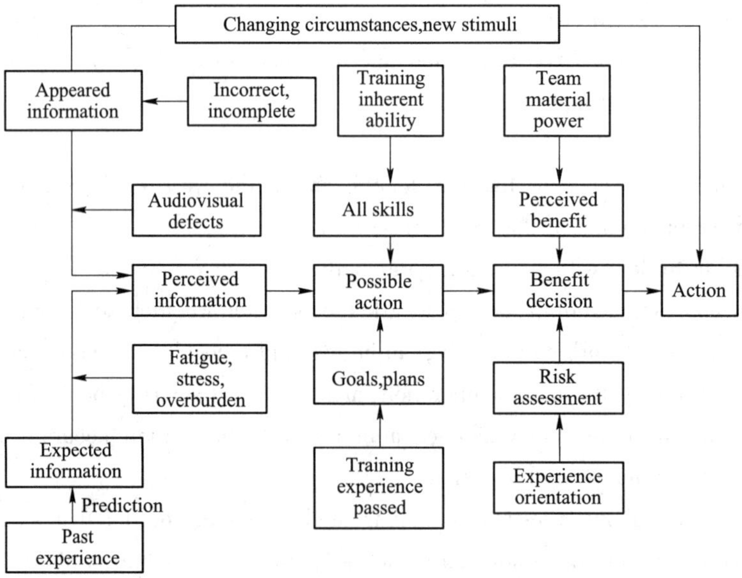

Figure 5-17 Haier accident model

Information includes information received by the operator in the operating system. This information may be incorrect due to mechanical failure, or imperceptible due to poor eyesight and hearing, that is, incomplete information, both of which may lead to actions Mistake.

Expected information refers to forecasts that often guide the collection and selection of information. In terms of predicting the guidance perception, two types of errors may occur: one is the operator's perception of the error, and the other is the unawareness of the danger signs. Only when the information shows that it is not safe, predictions can draw analogies.

When the burden is too heavy, there is pressure, fatigue, or the effects ofdrugs, the operator's attention to collecting information is weakened, so that it is impossible to maintain alertness to danger. Table 5-5 gives the brain level of conscionsness. Table 5-6 gives an investigation of drivers' physiological changes.

Table 5-5 Brain level of consciousness

Status level	Attention status	Attention	Physiological state	Reliability
0	Unconsciousness or loss of sanity	Zero	Sleep or some kind of brain disease	0
1	Abnormal consciousness	Slow attention	Fatigue, falling asleep, monotonous, drunk	≤0.9
2	Relaxation phase of normal consciousness	Passive, absent-minded	Quiet, rest, routine work	0.99~0.999999
3	The waking phase of normal consciousness	Proactive and wide-ranging	Active	>0.999999
4	Very long or extremely excited	Condense a certain point, judge to stop	Panic, nervousness, etc. during emergency defense response	≤0.9

Table 5-6 Investigation of drivers' physiological changes

Status	Heart rate (beats/min)	Eye movement (times/second)	Visual distance (m)
While driving	83	0.85	92.5
When not driving	72	0.65	110

Behavioral decision-making: Based on the perceived information, after processing, whether the correct action can be taken depends on the guidance, training and inherent ability. Decision-making should consider economic and social benefits, including the interests of production teams and groups, as well as the original experience and the subjective assessment of dangers that result from it.

Recognition, understanding, and decision-making all belong to central processing, followed by actionoutput (response behavior). After the response action, the operating system will change. Inspection and monitoring functions are the main functions in the feedback loop.

5.7 The origin of disturbance

5.7.1 Disturbance origin accident model

In 1972, Benner proposed a comprehensive concept and terminology to explain the causes of accidents. At the same time, the branch event chain and the accident process chain were combined and displayed with a logical diagram.

He pointed out that starting from the purpose of investigating the cause of the accident, treating an event as something that has happened is an instantaneous major change in the situation and an accident that leads to the next event. The occurrence of an event is bound to be caused by related persons or things.

The relevant persons or things are collectively called "actors", and their behaviors are called "behaviors". In this way, an event can be described by the terms "actor" and "behavior". The "actor" can be any living body, such as a car worker, driver, or factory manager; or any non-living substance, such as machinery, wheels, or design drawings. "Behavior" can be anything that happens, such as movement, malfunction, observation, or decision-making. The incident must be described in terms of individual actors and behaviors, so that the accident process can be broken down into several parts for analysis and synthesis.

In 1974, Lawrence used the above theory to put forward the theory of the origin of disturbance. The theory holds that "events" are factors that constitute accidents. Any accident has some kind of abnormal "disturbance" when it is in its infancy, and this disturbance is the origin event. The accident formation process is a group of consciously or unconsciously, pointing to some kind of expected or unexpected result. This kind of accident process includes the influence of external conditions and their changes. The process of successive events is carried out in an automatically adjusted dynamic balance.

If the actor behaves properly or the force is moderate, the energy flow can be maintained stable without deviation, thereby achieving safe production. If the actor behaves improperly or has a failure, the above balance will be disturbed, which will destroy and end the automatic dynamic balance. When the accident process begins, one event is followed by another event, which eventually leads to the "end event" accidents and injuries. Such accidents and injuries or damages will in turn cause energy release or other changes.

Disturbance origin theory regards accidents as perturbation in the course of successive events and ending in injury or damage. This can be called "P Theory" (Perturbation Theory). P Theory general accident model is shown in Figure 5-18.

The periphery of the graph is automatically balanced, without the consequences of accidents, only making production activities abnormal. The figure also shows that at the time of the incident, if conditions are improved, the chain of events can also be interrupted, preventing the progress of the incident from developing and turning into safety.

Event terms are highly abstract "stress" terms to adapt to various states.

5.7.2 Multilinearity and application of accident event process

The diagram of the process of multiple linear events can be analyzed according to the sequence of events, related factors of the accident and the relationship with other events.

When combined with the above model provided by P Theory, it is a more effective tool for investigating and analyzing accidents.

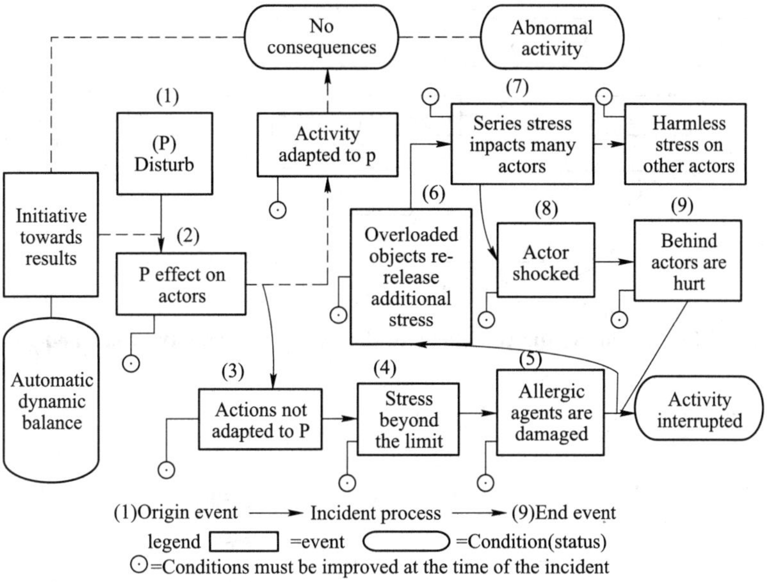

Figure 5-18 P Theory general accident model

Like most system safety analysis, rectangular boxes are used here to represent events, and ellipses are used to represent conditions.

Figure 5-19 shows the events that constitute an activity and the results of such an activity for an actor.

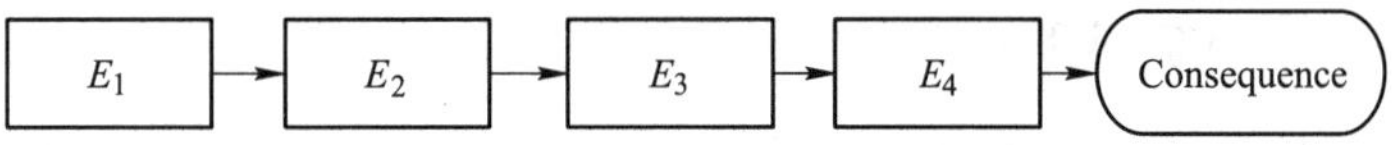

Figure 5-19 The events that constitute an activity and the results of such an activity for an actor

When two or more actors produce results, it is shown in Figure 5-20.

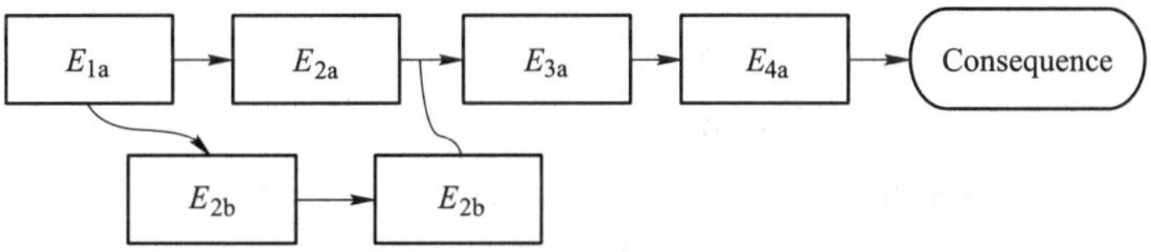

Figure 5-20 Two or more actors produce results

The interval of each event in the figure can be used to indicate the timing of the event relative to other events.

The arrows indicate the flow relationship of events or the logical relationship before and after the event, and can also indicate conditions similarly. Refer to Figure 5-21. The chronological and logical sequence of events occurring in the accident process is shown in Figure 5-22. An actor's condition and event separation procedure is shown in Figure 5-23.

The use of P Theory and diagrams can strengthen the explanation of accident phenomena and help overcome the weaknesses of other accident models.

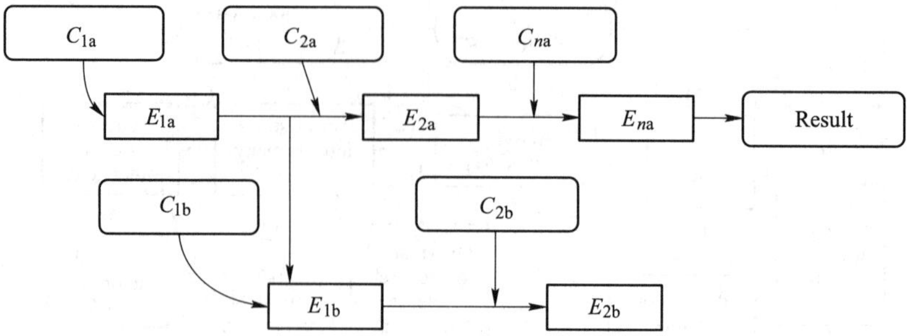

Figure 5-21 Including the conditions of the two actors activity events and results

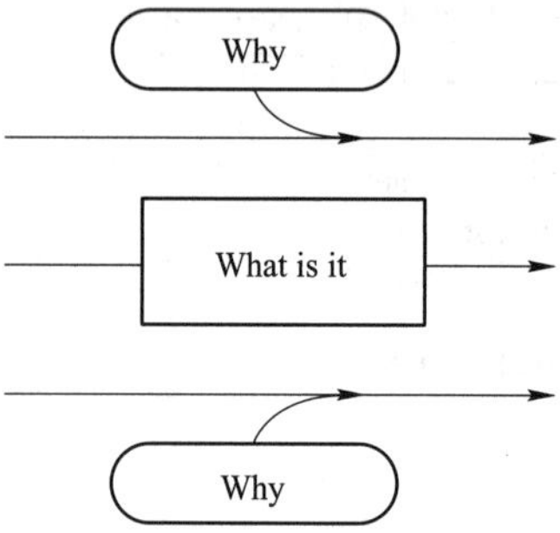

Figure 5-22 Description of the simple method

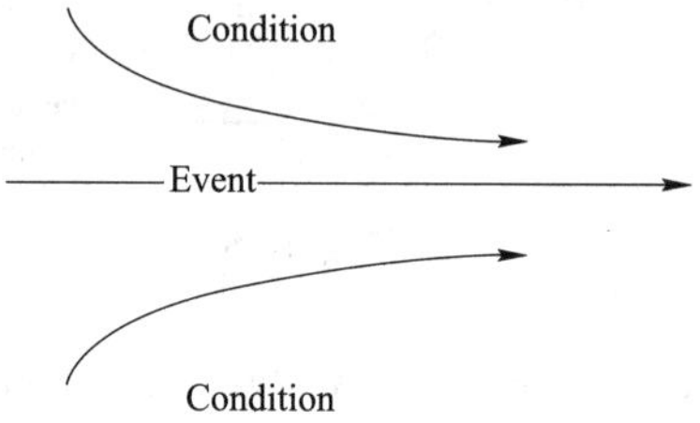

Figure 5-23 An actor's condition and event separation procedure

5.8 Dynamic change theory

Changes in states and elements will have a substantial impact on most systems. Studying the change of a certain part and its impact on safety, what kind of result it can lead to high-level subsystems and the whole system, is one of the most basic tasks of system safety analysis.

When studying and analyzing accidents, the "changes" in the system and the "errors" caused by the changes must be considered as a basic element. When a production process or operation is out of control, changes will obviously occur.

Changes include: (1) expected and planned changes; (2) unexpected changes. Most accident causes involve changes. Therefore, changes will lead to accidents. At the same time, changes can also be used to create some safety conditions. "Change" can also be used as a method of judging the cause and effect of an event. Therefore, "change" should be studied as a basis for evaluating the possibility of accidents.

During the production process, the enterprise equipment is updated, and the process and technology are constantly changing. In response to changes in objective reality, accident prevention should be improved accordingly to adapt to the changed situation. If the manager cannot or fails to adapt to the change in time, management errors will occur; if the operator cannot or fails to adapt to the change in time, operating errors will occur; changes in external conditions can also cause failures of machinery and equipment, which may lead to accidents.

Johnson paid attention to the role of change in the occurrence and development of accidents early on. He defined an accident as an undesirable or unexpected release of energy, which occurred due to a manager's planning error or operator's misbehavior, and then did not adapt to changes in material or human factors in the production process, resulting in unsafe. The behavior or unsafe state destroys the shielding or control of energy, causing casualties or property losses during the production process. Figure 5-24 shows Johnson's accident causal chain model.

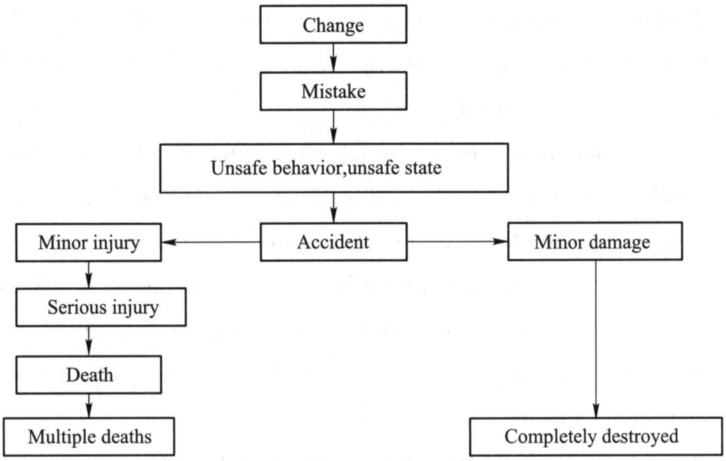

Figure 5-24 Johnson's change-mistakes theory

In the work of safety production, changes are regarded as a potential cause of accidents, which should be discovered as early as possible and corresponding measures should be taken. As a safety manager, you should pay attention to the following changes:

(1) Changes outside the enterprise and changes within the enterprise. The social environment outside the enterprise, especially the changes in national politics, economic policies, and policies, has a huge impact on the internal management and personnel thinking of the enterprise. For example, looking at the situation of industrial injuries after the founding of New China, it can be found that during the two major social changes during the Great Leap Forward and the Cultural Revolution, the internal order of enterprises was disrupted, and injury accidents increased significantly. In response to changes outside the company, the company must take appropriate measures to adapt to these changes.

(2) Macroscopic changes and microscopic changes. Macro changes refer to changes in the enterprise as a whole, such as the replacement of leaders, new employees seeking employment, personnel adjustments, and changes in production conditions. Micro changes refer to changes in specific things. Through micro changes, safety managers should discover hidden problems behind them and take appropriate countermeasures in time.

(3) Changes within and outside the plan. For planned changes, hazard analysis should be conducted in advance and safety measures should be taken. For unplanned changes, first the changes are found, and then improvement measures are taken based on the changes found.

(4) Actual changes and potential or possible changes. The actual changes can be found through observation and inspection. The discovery of potential or possible changes requires analysis and research.

(5) Changes in time. Over time, performance is low or degraded, and interacts with other changes.

(6) Changes in technology. People make mistakes because they are unfamiliar with new processes, new technologies or starting new engineering projects.

(7) Changes in personnel. Changes in all aspects of personnel affect people's work ability, causing operational errors and unsafe behavior.

(8) Changes in labor organization. Changes in labor organization and poor shifts have resulted in unconnected work, which in turn leads to human error and unsafe behavior.

(9) Changes in operating procedures. It should be noted that not all changes are harmful. The key lies in whether people can adapt to changes in objective conditions. In addition, changes are often used in accident prevention to prevent human errors. For example, pipes of different colors are used to transport different gases as required; the operating handles and buttons are made into different shapes to prevent confusion.

When applying the viewpoint of change for accident analysis, the changes can be found by the difference between the current state and the previous state of the following factors: (1) objects, protective devices, energy, etc; (2) personnel; (3) tasks, goals, procedures, etc; (4) working conditions, environment, time arrangement, etc; (5) management work, supervision and inspection, etc.

Johnson believes that accidents are often caused by multiple reasons, including a series of changes and mistakes. For example, the mistakes of business leaders, planners, supervisors, and operators are shown in Figure 5-25.

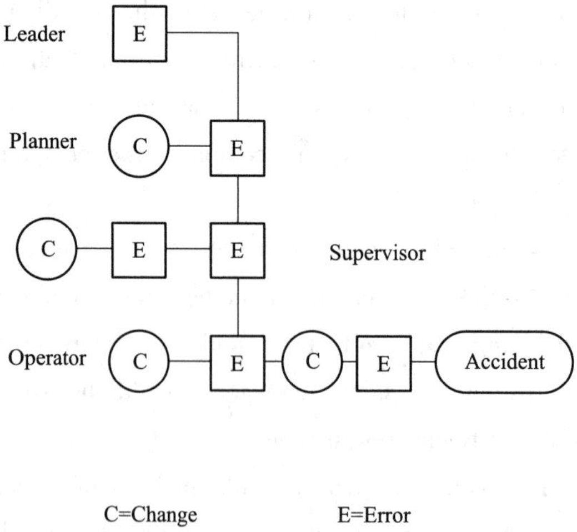

Figure 5-25 Change-fault linkage model

It can be seen from the Figure 5-26 that, starting from welding defects, a series of changes and mistakes have resulted in a fire accident in the gas pipeline.

Change-fault theory:

(1) Change 1: a set of larger and more efficient equipment was introduced.

(2) Change 2: the old equipment was disabled and part of it was dismantled.

(3) Change 3: the operation and efficiency of the new equipment did not meet the requirements at first.

(4) Change 4: the demand for the factory's products has increased dramatically.

(5) Change 5: re-enable the old device.

(6) Change 6: requires that the operation control of the old equipment be restored and put into productionas soon as possible.

(7) Mistake 1: not doing a comprehensive risk analysis and adequate preparations.

(8) Variation 7: some redundant safety control devices are not activated immediately.

(9) Result: the equipment exploded and 6 people died.

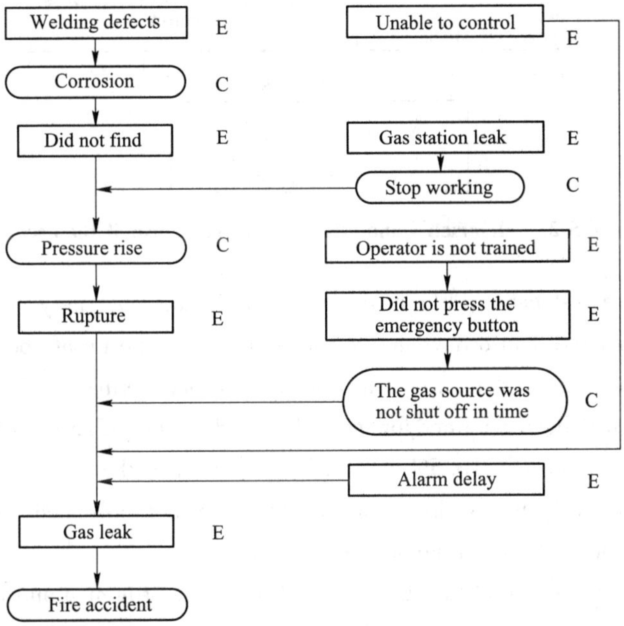

Figure 5-26 The change and error analysis of the accident caused by the rupture of the gas pipeline

5.9 Trajectory intersection theory

5.9.1 The position of people and things in the cause of the accident

The unsafe behavior of people and the unsafe state of things are the direct causes of industrial injury accidents. The recognition of the status of unsafe human behavior and unsafe conditions in the cause of accidents is an important issue in the theory of accident causes.

According to Heinrich's research, the main cause of the accident was either due to unsafe behavior of people or unsafe state of things. No accident was caused by both unsafe behavior of people and unsafe state of things (See Figure 5-27) . Therefore, he concluded that almost all industrial injury accidents are caused by human unsafe behavior. Later, this view was criticized by many researchers.

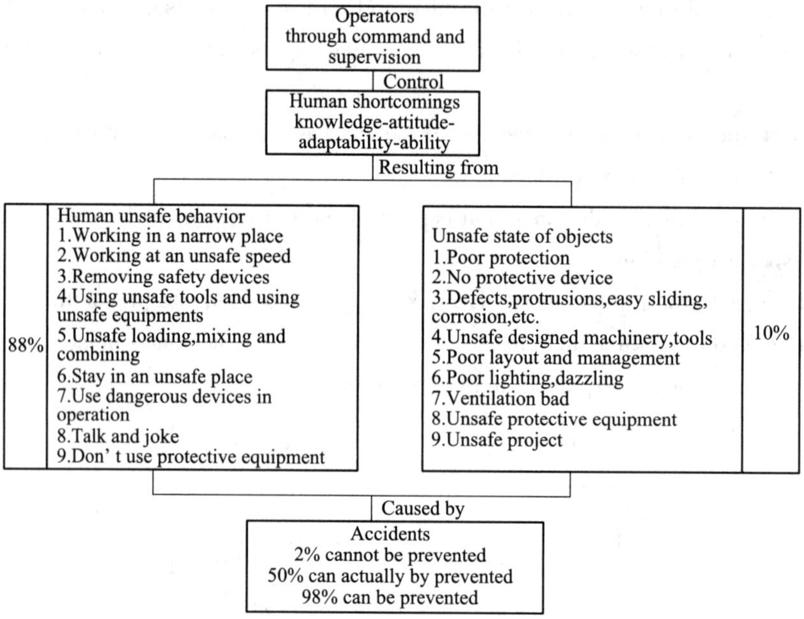

Figure 5-27 Heinrich's analysis of the direct cause of the accident

According to Japanese statistics, 96% of the accidents in the machinery manufacturing industry that were off work for more than 8 days in 1969 were related to the unsafe behavior of people, and 91% were related to the unsafe state of objects; machinery manufacturing in 1977. Among the 104 638 injury accidents that had been off work for more than 4 days, only 5. 5% had nothing to do with human unsafe behavior, and only 16. 5% had nothing to do with the unsafe state of things. These statistics show that most industrial injury accidents occur not only because of human unsafe behavior, but also because of the unsafe state of things.

The change in the recognition of the status of the two factors of human and material in the cause of accidents is due to the advancement of production technology, and the problem of unsafe production equipment and production conditions has attracted more and more attention. The in-depth study of human factors can correctly distinguish between unsafe human behavior and unsafe state of things.

As Johnson pointed out, judging whether it is an unsafe behavior or an unsafe state is influenced by the researcher's subjective factors and depends on his deep understanding of the problem. Many people, due to lack of knowledge about human error, regard the unsafe state caused by human error as unsafe behavior. Now, more and more people realize that the reason why an industrial accident can happen is that in addition to human unsafe behavior, there must be some unsafe conditions.

R. Skiba pointed out that both production operators and machinery and equipment have an impact on the occurrence of accidents, and the dangerous state of machinery and equipment has a greater effect on the occurrence of accidents.

He believes that only when two factors appear at the same time can an accident occur. Practice has proved that eliminating unsafe conditions in production operations can greatly reduce the occurrence of injury accidents. For example, before the installation of automatic connectors on railroad vehicles in the United States, hundreds of railroad workers died in vehicle connection accidents every year. The head of the railway department attributed the responsibility for the accident to the worker's error or carelessness. Later, in accordance with the requirements of the government decree, all railway vehicles were equipped with automatic connectors. As a result, fatal accidents during the vehicle connection operation were greatly reduced.

5.9.2 Trajectory intersection theory and accident cause model

Figure 5-28 shows the trajectory in tersection theory accident model. Specifically, the factors of the two event chain of people and things are as follows.

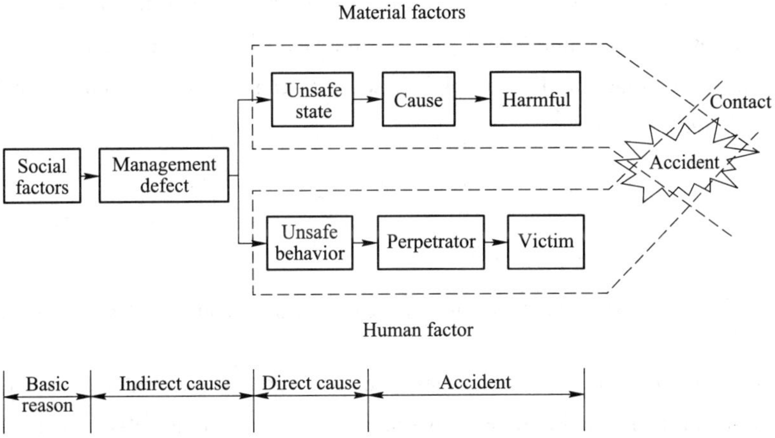

Figure 5-28 Trajectory intersection theory accident model

(1) Human event chain Human unsafe behavior is based on physical, psychological, environmental, and behavioral aspects:

1) Physiological inheritance, congenital heart defects;

2) Defects in social environment and business management;

3) Acquired psychological defects;

4) Sensory differences in sight, hearing, smell, taste, touch, etc.;

5) Behavioral errors.

People have great freedom of action, and they are affected by environmental conditions during production and labor, coupled with their own physical and psychological defects, are prone tomisoperations or behavioral errors.

The movement trajectory of the human event chain over time is carried out in the order of ①→②→③→④→⑤.

(2) Event chain of objects in the machinery and material series, from the beginning of design, through various procedures on the spot, unsafe conditions may occur at all stages of the entire production process.

1) Defects in design and manufacturing, such as improper materials, incorrect strength calculations, poor structural integrity, wrong processing methods or low processing accuracy;

2) Defects in the process flow, such as the mining method is not suitable for the nature of the surrounding rock of the deposit;

3) Defects in maintenance, which reduce reliability, such as equipment wear, aging, overload operation, poor maintenance, etc.;

4) Defects in operation;

5) Defects in the workplace environment.

The trajectory of the event chain of matter or machinery along the time course is carried out according to the direction line of A→B→C→D→E.

The intersection of the movement trajectory of the human factor chain and the movement trajectory of the thing factor chain, that is, when the unsafe behavior of human and the unsafe state of the thing appear at the same time, accidents and injuries will occur. The time and place at which the two event chains of human and material intersect are the "time and space" of casualties. As shown in Figure 5-29.

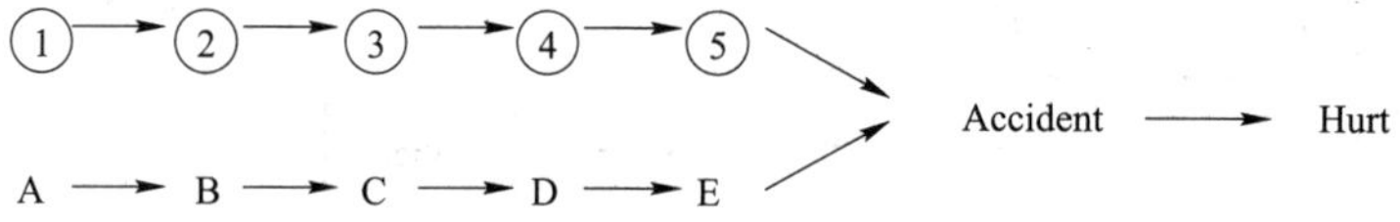

Figure 5-29 The relationship between two event chains of human and material intersect

In most cases, due to poor enterprise management, workers lack education and training, or mechanical equipment lacks maintenance, overhaul, and incomplete safety devices, resulting in unsafe human behavior or unsafe conditions. If we try to eliminate the hidden dangers of mechanical equipment or the process of handling dangerous substances, or eliminate human error and unsafe behavior, so that the two event chains are interrupted, the two series of motion trajectories cannot intersect, and the danger will not appear, and safe production can be achieved.

According to the view of trajectory intersection theory, eliminating unsafe behaviors of people can avoid accidents. Emphasizing the assessment of work types, strengthening safety education and technical training, conducting scientific safety management, and controlling the occurrence of unsafe actions by people from physical, psychological and operational management is equivalent to cutting off the human factor trajectory of the accident.

However, it should be noted that humans are different from mechanical equipment. Machines operate under the constraints prescribed by people and have less freedom; while human behavior is governed by their own thoughts and has greater freedom of behavior.

On the one hand, this freedom of behavior makes people have the initiative to do a good job of safe production. On the other hand, it may also cause people's behavior to deviate from predetermined goals and cause unsafe behavior. Because human behavior is affected by many factors, controlling human behavior is a very difficult task.

The unsafe state of elimination can also avoid accidents. By improving the production process and setting up effective safety protection devices, the dangerous conditions in the production process are eradicated, so that even if personnel produce unsafe behaviors, accidents will not be caused.

In safety engineering, the safety of production conditions such as mechanical equipment and physical environment is called intrinsic safety. In all safety measures, the first thing that should be considered is to realize the intrinsic safety of the production process and production conditions. Practice has proved that eliminating unsafe conditions in production operations can greatly reduce the occurrence of casualties.

The focus of the trajectory intersection theory is to explain that human errors are difficult to control, but the controllable equipment and logistics will not fail. Some managers, and even some leading cadres, always mistakenly attribute all casualties to the operators' "violating operation". In fact, the unsafe actions of people are also caused by insufficient education and training and other management deficiencies.

The focus of management should be on the unsafe state of the controlled objects, that is, to eliminate the "causal objects", and of course there will be no "harmful objects" and "cut off" the logistics chain event chain, so that the flow of people and the trajectory of logistics do not cross. The accident can be avoided.

For example, for the human series, strengthening the assessment of chemical species, strengthening safety education and technical training, conducting scientific safety management, and controlling the occurrence of unsafe behaviors from physical, psychological, and operational management is equivalent to cutting off human the chain of events. However, as mentioned earlier, it is difficult to control people who have a large degree of freedom and large differences in physical and mental personality, and accidental mistakes are hard to avoid.

The trajectory intersection theory emphasizes the event chain of cutting off objects, advocates the use of systems and equipment with high reliability and strong structural integrity, and vigorously promotes insurance systems, protection systems, signal systems, and automatic remote control devices. This way, even if human error, The ①→⑤ series will also control the development of the A→E series due to the role of high-reliability safety systems such as safety locks, which can completely avoid casualties.

Therefore, in safety engineering, the first thing to consider is to realize the essential safety of the production process and production conditions, that is, the mechanical equipment, materials and environment. Set up effective safety protection devices, even if personnel work and operating errors will not cause accidents. However, even when safety technical measures have been adopted, safety protection devices have been added to reduce and control the unsafe state of logistics, safety education, safety training, and consultation on safety psychology for workers and cadres must still be strengthened. Strictly implement safety regulations and operation standardization to regulate human behavior and prevent human errors.

In short, according to the viewpoint of trajectory intersection theory, in order to effectively prevent accidents, measures must be taken to eliminate unsafe behaviors of people and unsafe conditions of objects.

5.10 Synthesis

It can be seen from the analysis of the above-mentioned various accident cause theories that the unsafe behavior of people and the unsafe state of things are the direct causes of accidents. If they are further considered, two kinds of Deep-seated reasons behind the author.

Nowadays, safety experts at home and abroad generally believe that accidents are not caused by a single factor, nor are they caused by accidental personal errors or simple equipment failures, but the result of a combination of various factors.

Comprehensive theory believes that the occurrence of accidents is the result of social factors, management factors, and various hazards in production triggered by accidents. Comprehensive theory believes that the timely process of an accident is a disaster phenomenon formed by the cause and the perpetrator accidentally triggering the harm and the victim. Comprehensive accident model is displayed in Figure 5-30.

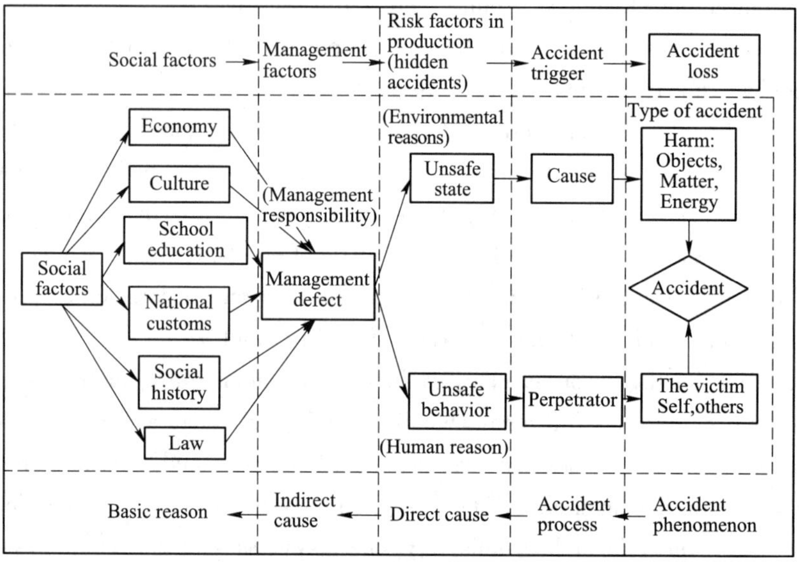

Figure 5-30 Comprehensive accident model

The accidental event is triggered because the environmental conditions in the production have various hidden dangers (the unsafe state of objects) and some human errors (human unsafe behavior) which together constitute the direct cause of the accident.

These material, environmental, and human causes are caused by management errors, management defects, and management responsibilities. This is an indirect cause that forms a direct cause and an important basic cause. The factors that form indirect causes, including basic social factors such as economy, culture, education, habits, history, and laws, are collectively referred to as social factors.

Obviously, this theory comprehensively considers various accident phenomena and factors, so it is relatively correct and is conducive to the analysis, prevention and handling of various accidents. It is the most popular theory in the world today. The United States, Japan, and China all advocate analyzing accidents in this way.

The process of accidents can be expressed as the basic causes of "social factors" to produce "management factors", and further to produce "risk factors in production". Casualties and losses occur through accidental factors of people and things.

The process of investigating and analyzing the accident is the opposite of the above experience. For example, reverse tracking inquire about the accident process through the phenomenon of accidents and then understand the environmental reasons and human reasons that directly caused the accident. Investigate management responsibilities (indirect causes) and social factors (basic causes) in turn.

5. 11 Modern domestic scholars and theories

5. 11. 1 Bird theory

Bird accident causal chain is a modern accident causal chain theory proposed by Frank Bird on the basis of Heinrich accident causal chain theory, see Figure 5-31. It believes that the direct cause of the accident is the unsafe behavior of people and the unsafe state of things. The indirect cause include personal and work-related factors. The root cause is the management defect. The problems or defects in management are the reasons for the existence of indirect causes. The existence of indirect causes leads to the existence of direct causes, which ultimately leads to the occurrence of accidents.

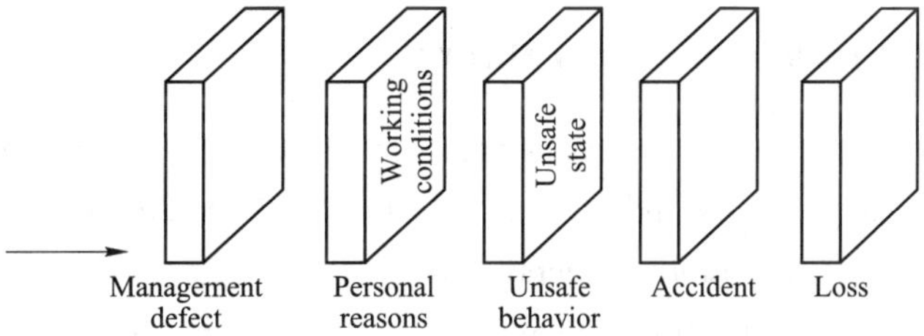

Figure 5-31 Bird accident causal chain theory

Bird accident causal chain process has five factors, but the meaning of each factor is different from Heinrich's. For most enterprises, due to various reasons, it is neither economic nor realistic to completely rely on engineering and technical measures to prevent accidents. Improving safety management and making great efforts can prevent accidents. Enterprise managers must realize that only by realizing the intrinsic safety of production, the accidents and injuries will not occur. Therefore, safety management is an important part of enterprise management. The safety management system should be constantly adjusted and improved with the development and change of production. A perfect management system cannot exist.

Due to the defects in safety management, other causes that can cause accidents appear. Personal reasons include lack of safety knowledge or skills, incorrect behavior motivation, physical or psychological problems, etc. The working conditions reasons include imperfect safety operation procedures, inappropriate equipment and materials, the existence of temperature, humidity, dust, gas, noise, lighting, working site conditions (such as slippery ground, obstacles, unreliable supports) and other harmful operating environmental factors. Only by finding out and controlling these causes can we effectively prevent the occurrence of subsequent causes, thus preventing the occurrence of accidents.

The direct cause of the accident is the unsafe behavior of people or the unsafe state of things. This is the reason that must be focused on in safety management. However, the direct cause is only a superficial phenomenon. It is the representation of the deep-seated cause. In actual work, we should not stay on the surface. We should investigate the causes of hidden management defects and take effective control measures to fundamentally eliminate the occurrence of accidents.

The accident here is regarded as the contact of human body or object with energy exceeding the threshold, or the contact of human body with substances that interfere with normal physiological activities. Therefore, prevent accidents means prevent contact. The release of energy can be prevented by improving the devices, materials, processes, etc. The operator can improve their ability to identify and avoid hazards and wear personal protective equipment to prevent contact.

Personal injury and property damage are collectively referred to as loss. Personal injuries include industrial injuries, occupational diseases, mental trauma, etc. In many cases, appropriate measures can be taken to minimize the loss caused by accidents. For example, rescue the injured quickly and correctly, repair the equipment, conduct emergency training for relevant personnel at ordinary times, etc.

5.11.2 Accidental release of energy theory

Any industrial production process is a process of energy conversion or work. According to the accidental release of energy theory, industrial accidents and the injury or damage usually occur in the production process of uncontrolled energy conversion and (or) energy work. The accidental release of energy theory, starting from the physical nature of the accident, expounds the chain process of the accident. The unsafe behavior of people, the unsafe state of things and their interaction is caused by management mistakes, which releases abnormal or unwanted dangerous substances and energy. And they are transferred to human bodies and facilities, which causes casualties and (or) property losses. Accidents can be prevented by reducing energy and strengthening shielding. If different kinds of energy indispensable to human production and life lose control for some reason, it will be accidentally released or escaped against human will, which will stop ongoing activities and cause accidents, resulting in personal injury or property loss. Accidental release of energy theory is shown in Figure 5-32.

It is precisely because of the development, change and interaction of hazard sources in the

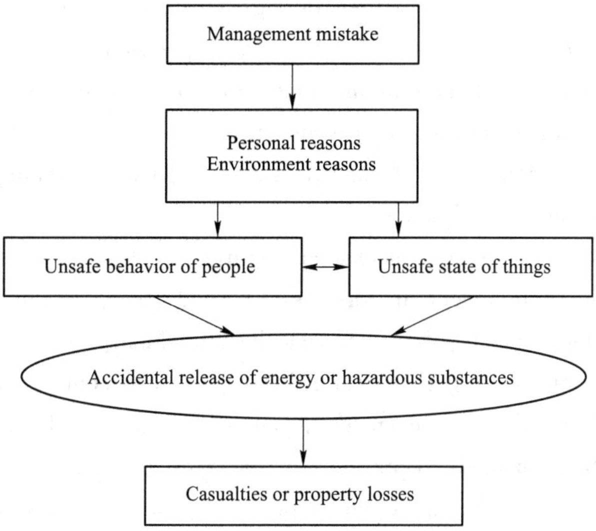

Figure 5-32 Accidental release of energy theory

system that energy is released unexpectedly. According to the role of hazard sources in the occurrence and development of accidents, they can be divided into two categories. The classification makes the object of accident prevention and control clearer. The first type of hazard source is different kinds of energy or hazardous substances that may be accidentally released in the system. The second type of hazard source is a variety of unsafe factors that lead to the failure or destruction of restraint and energy limiting measures. The existence of the first type of hazard is the premise of the accident. The second type of hazard source is the necessary condition for the first type of hazard source to cause accidents. The two types of hazard sources jointly determine the hazard of the hazard source. The energy released by the first type of hazard source is the energy subject that causes personal injury or property damage, which determines the severity of the accident consequences. The difficulty of the occurrence of the second type of hazard source determines the possibility of the accident.

In specific safety projects, the first type of hazard sources have objectively existed. And necessary measures have been taken to control them during design and construction. Their quantity and status are usually difficult to change. Therefore, the focus of accident prevention work is the second type of hazard sources. The focus of accident control is the first type of hazard sources.

Industrial safety engineering includes five parts.

(1) Chemical safety engineering. Chemical safety engineering technicians shall understand the mechanism of fire, explosion and poisoning accidents caused by chemical energy release. Understand the hazards of chemical substances. Master the analysis methods of general chemical fire and explosion accidents, and be able to put forward corresponding preventive measures. Master the prevention and control methods of chemical fire and explosion accidents and chemical poisoning accidents.

(2) Building safety engineering. Building safety engineering technicians shall master the types of

common accidents and preventive measures during construction, and focus on the causes of accidents and preventive measures for falling from high places. Be familiar with the main risk factors affecting the building body and the main measures for accident prevention. Be familiar with the main contents of building fire protection engineering.

(3) Mechanical safety engineering. Mechanical safety engineering technicians shall understand the basic knowledge of mechanical safety design, devices and use. Grasp the types of mechanical injuries and the measures to prevent mechanical injuries, as well as the safety technology and measures of mechanical equipment in the design, production, use, maintenance and other aspects. In particular, they shall be able to prevent the lifting and pressure vessels accidents.

(4) Electrical safety engineering. Electrical safety engineering technicians shall understand the general situation of electrical safety technology, types of electrical accidents, basic knowledge of insulation protection, basic methods to prevent electric shock, characteristics and hazards of static electricity, basic parameters of lightning, and types and hazards of lightning. Master the barrier protection and spacing protection of electrical safety, protective grounding and protective neutral connection of electrical equipment and facilities, safety power supply technology, electrical fire and explosion prevention measures, basic ways to eliminate static electricity, lightning protection classification and devices, and basic lightning protection technology.

(5) Occupational hazards and man-machine safety engineering. Safety engineering technicians shall master the sources, types and protective measures of occupational hazards. Understand the purpose of workplace environmental monitoring and the occupational health requirements of industrial sites, and master the control measures for industrial toxic hazards. Understand the main research contents of man-machine safety engineering, and master the common accidents of man-machine system and their causes.

5.11.3 Safety system theory

Safety system theory studies safety management from a systematic perspective. The main research objects are accident system and safety system. The four elements of the accident system are unsafe behavior of men, unsafe state of machine, poor medium factors, and inadequate management measures. It is usually called "4M" element. The safety system involves four elements: men-men safety quality (psychological and physiological, safety ability, cultural quality), material-safety and reliability of equipment and environment (design safety, manufacturing safety, use safety), energy-safety role of production process energy (effective control of energy), and information-full and reliable safety information flow (full play of management efficiency), which is the basic guarantee of safety. The system safety analysis developed from this is to explore the interaction, feedback and adjustment between men, machine and environment. And to find out the risk factors in the system, identify the state characteristics of the system when the accident is about to occur, especially the process characteristics related to men feeling, memory, understanding, behavior response, etc. It helps to distinguish the primary and secondary causes of the accident, so as to take corresponding measures to eliminate the system fault or accident. It can be widely used in the key decomposition of the tourism safety accident system and the element description and construction of the tourism safety management system. Safety system theory includes many innovative concepts different from traditional security theory.

(1) In the aspect of accident cause theory, it has changed the traditional concept that people only pay attention to the unsafe behavior of operators andignore the role of hardware failures in the accident cause. People started to consider how to improve the safety of complex systems by improving the reliability of the system, so as to avoid accidents.

(2) Nothing is absolutely safe. Danger factors hides in everything. Generally speaking, safety or danger is just a subjective judgment. The potential risk factors that can cause accidents are called hazard sources, and the possibility of causing personal injury or material loss from certain hazard sources is called danger. Hazard sources are some things or environmental factors that may cause problems, while hazards represent the opportunity of injury or loss caused by potential hazard sources, which can be measured by probability.

(3) It is impossible to eliminate all hazard sources and hazards. But it is possible to reduce the risk from existing hazard sources. The overall risk should be reduced instead of eliminating only a few selected hazards.

(4) Because people have limited cognitive ability, they sometimes cannot fully understand the hazard sources and hazards. Even if they know the existing hazard sources, new hazard sources will be generated with the development of production technology and the emergence of new technologies, processes, materials and energies. Due to the limitations of technology, capital, labor and other factors, it is impossible to completely eradicate the recognized hazard sources. Because the hazard sources cannot be completely eradicated, the hazard can only be reduced to an acceptable level, which is an acceptable hazard. The goal of safety work is to control the hazard sources, try to minimize the probability of accidents, and in case of accidents, control the injuries and losses to a lesser extent.

5.11.4 Trace intersecting theory

Trace intersection theory studies the causes of casualty accidents. It can be summarized as follows. When trace intersection of the two event chains: equipment failure (or unsafe state of things) and human error, an accident will occur. In most cases, due to poor enterprise management, lack of education and training of workers, or lack of maintenance, overhaul of machinery and equipment, as well as incomplete safety devices, unsafe behavior of people or unsafe state of things are caused. Trace intersection theory describes the occurrence and development process of accidents as follows: basic cause→indirect cause→direct cause→accident→injury. From the perspective of accident development movement, such process is described as the movement track of accidents caused by accident factors, including the movement trace of human factors and things factors. Figure 5-33 shows trace intersection theory.

The movement trace of human factors: unsafe behaviors of human are based on physiology, psychology, environment and behavior.

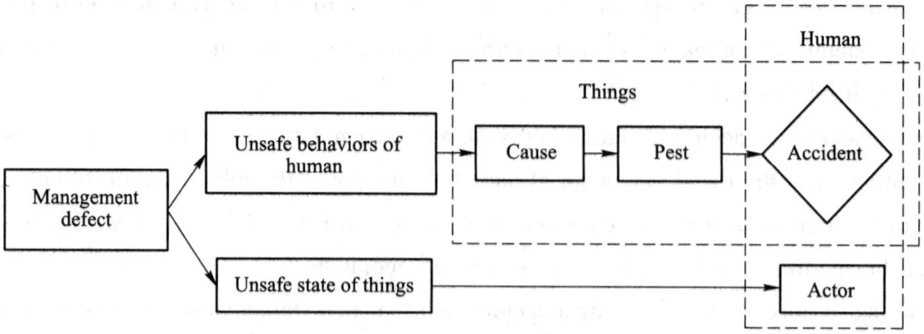

Figure 5-33 Trace intersection theory

(1) Physiological and congenital physical and mental defects.

(2) Social environment and enterprise management defects.

(3) Acquired psychological defects.

(4) Difference in sensory energy distribution of vision, hearing, smell, taste and touch.

(5) Behavior error.

The movement trace of things factors: unsafe state may occur in all stages of the production process.

(1) Defects in designs, such as improper use of materials, wrong strength calculation, poor structural integrity, mining method not suitable for the nature of the surrounding rock of the deposit, etc.

(2) Defects in manufacturing and process flow.

(3) Defects in maintenancewhich reduce reliability.

(4) Defects in use.

(5) Defects in the workplace environment.

In the production process, the movement trace of human factors shall be carried out in the direction of (1)→(2)→(3)→(4)→(5), and the movement trace of things shall be carried out in the direction of (1)→(2)→(3)→(4)→(5). The time and place where the two traces of human and things intersect is the "time and space" when casualty accidents occur, which also leads to the occurrence of accidents. It is worth noting that in many cases, human and things are mutually causal. For example, sometimes the unsafe state of things induces unsafe behaviors of human, and unsafe behaviors of human promote the development of the unsafe state of things or lead to the emergence of new unsafe states. Therefore, the actual accident does not simply follow the above two tracks of human and things, but presents a very complex causal relationship.

If we try to eliminate the hidden dangers behind mechanical equipment or in the process of handling dangerous substances, or try to eliminate human errors and unsafe behaviors, the chain of two events can be interrupted, the two series of motion traces cannot intersect, the danger cannot occur, and the accident can be avoided. For human factors, emphasizing the assessment of work types, strengthening safety education and technical training, conducting scientific safety management, and controlling the occurrence of unsafe behaviors of people from the aspects of physiology, psychology and operation management are equivalent to cutting off the trace of human factors caused by accidents. However, it is difficult to control people with large degrees of freedom and large differences in physical and mental temperament, and it is difficult to avoid accidental mistakes. In most cases, due to poor enterprise management, workers lack education and training, or mechanical equipment lacks maintenance and repair, or safety devices are incomplete, resulting in unsafe behavior of human or unsafe state of things.

Trace intersection theory highlights cutting the movement trace of things factors, advocates the use of systems and equipment with high reliability and structural integrity. It vigorously promotes insurance systems, protection systems, signal systems, and highly automated and remote control devices. In this way, even if human error constitutes human factors (1) to (5), the function of safety locking and other highly reliable safety systems will also control the development of things factors (1) to (5), which can completely avoid casualty accidents. Some leaders and managers always wrongly blame all casualty accidents on the "illegal operation" of operators. In fact, unsafe behaviors of human is also caused by lack of management, such as insufficient education and training. The key point of management should be set to control the unsafe state of things, that is, to eliminate the "cause". Then there will be no "pest". Cut the movement trace of things factors. So that the traces of human and things do not intersect, and accidents can be avoided. Practice has proved that eliminating the unsafe state of things in production can greatly reduce the occurrence of casualty accidents.

5.11.5 Kitagawa Chusan accident causal chain theory

Accident causal chain theories such as Heinrich causal chain theory, Bird causal chain theory and Adams causal chain theory, all limited the scope of investigation to the enterprise. Chusan Kitagawa of Japan believed that the causes of industrial injury accidents are very complex. Enterprises are a part of society. The political, economic, cultural, scientific and technological development, and many other social factors of a country or region have an important impact on the occurrence and prevention of internal injury accidents in enterprises. Therefore, chusan kitagawa proposed a new accident causal chain theory, as shown in Table 5-7.

Table 5-7 Kitagawa Chusan accident causal chain theory

Basic reasons	Indirect reasons	Direct reasons	Accident	Injure
School education reasons; Social and historical reasons; Management reasons	Technical reasons; Educational reasons; Physical reasons; Mental reasons	Unsafe behavior; Unsafe state		

Kitagawa Chusan believed that the basic reasons of the accident should include the following three aspects:

(1) Management reasons. Enterprise leaders do not pay enough attention to safety. So management defects appears, for example, the operation standard is not clear, the maintenance system is

defective, the personnel arrangement is improper, and the enthusiasm of employees is not high.

(2) School education reasons. Insufficient safety education in primary school, middle school, university and other educational institution.

(3) Social or historical reasons. The concept of social security is backward, the safety regulations or safety management or supervision institutions are incomplete, etc.

Indirect reasons:

(1) Technical reasons. Defects in design, construction, maintenance and other technical aspects of machinery, devices, buildings, etc.

(2) Educational reasons. Due to the lack of safety knowledge and operation experience, they do not know or despise the dangers and safe operation methods in the operation process, or they are unskilled and used to the operation.

(3) Physical causes. Poor physical condition, such as headache, coma, epilepsy and other diseases, or physiological defects such as myopia and deafness, or fatigue, lack of sleep, etc.

(4) Mental causes. Negative, resistant, discontented and other bad attitudes, anxiety, tension, fear, extreme and other mental instability, narrow-minded, stubborn and other bad characters, and intellectual obstacles.

Among the above four indirect causes, the first two are more common, and the last two are less common.

5.11.6 24Model

The English abbreviation of the accident causes 24 model is "24Model", as shown in Figure 5-34. It is a modern accident causes theoretical model proposed by the Safety Management Research Center of China University of Mining and Technology (Beijing) after 10 years of research. The main theoretical basis of 24Model is the domino series model proposed and improved by Heinrich and Bird, the Swiss cheese series model and human factor classification system proposed by Reason and Shappell, and the excellent safety management model proposed by Stewart. The logic structure of the static 24Model is based on the causal relationship, which belongs to the causal sequence accident cause model. The meaning of direct, indirect, root and basic cause of accidents in the static 24Model can be found from the above three categories of accident cause models, management system standards and the definition of safety culture. Figure 5-35 shows the dynamic structure, systematicity and behavior evolution process of 24Model. The dynamic 24Model establishes a logical structure based on the behavior evolution relationship, which fully expresses its own systematicness, dynamics and nonlinearity. And it forms a behavior evolution system by relying on organizational behavior which is influencing behavior, and individual behavior which is operating behavior. 24Model suggests that any accident occurs at least within the social organization. It divides the causes of the accident into the organization internal and external causes. The organization internal causes can be divided into two levels: organizational behavior and individual behavior, see Figure 5-36. Organizational behavior can be divided into two stages: safety culture (root cause) and safety management system (basic cause). Individual behavior can be divided into two stages: habitual behavior (indirect cause), one-time behavior and physical state (direct cause). There are four stages in total. The four stages are linked together to form a behavioral accident cause model. This is the origin of the name of "accident causes 24Model".

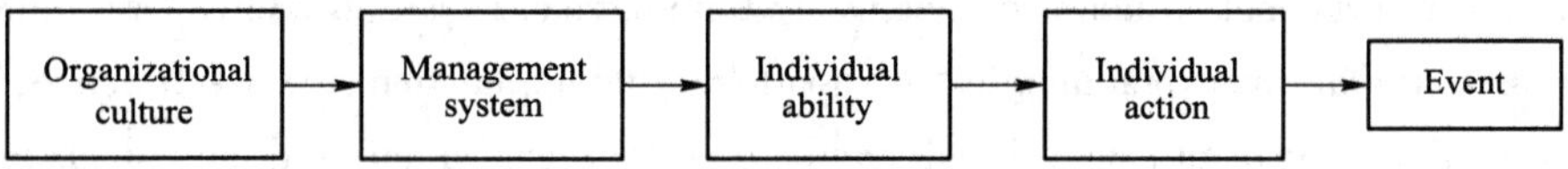

Figure 5-34 The static 24Model

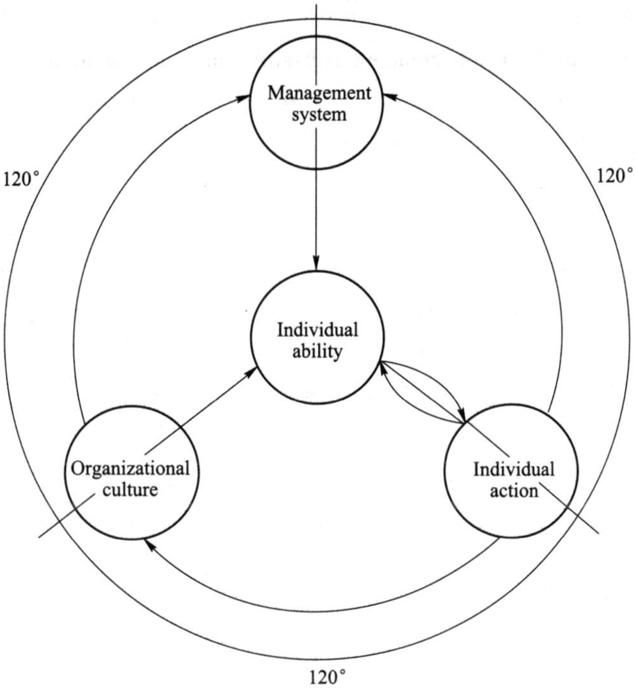

Figure 5-35 The dynamic structure, systematicity and behavior evolution process of 24Model

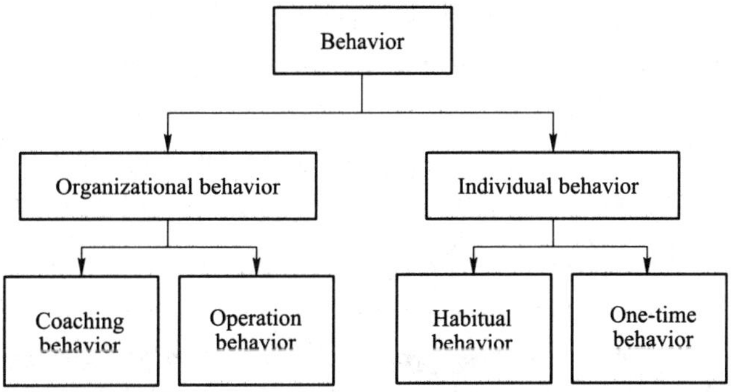

Figure 5-36 The behavior classification system of 24Model

The accident causes 24Model is not only a model for accident cause analysis, but also an accident prevention model for accident prevention countermeasure design. It can be used in the operation of accident prevention practices, such as the division of safety disciplines, laboratory planning and construction, personnel training program design in the field of safety, safety management organizational structure design, safety training, etc. It can also serve as the theoretical basis for safety management practice. In fact, apart from its application in safety science, 24Model is also a general management model, which can be used for organizations and individuals to manage anything.

[Exercises]

What is the guiding role of Heinrich accident causality chain in formulating accident prevention measures?

6 Safety Evaluation Model and Method

In recent years, safety accidents have occurred frequently, which is shocking and deplorable. On January 10, 2021, an explosion occurred in the return air shaft of Hushan Gold Mine in Qixia, Shandong, during the capital construction. It resulted in 10 deaths in the accident. The accident was a major production safety responsibility accident caused by the illegal storage and use of civil explosives and the illegal hot work at the wellhead. On February 17, a fire broke out in Caojiawa Gold Mine in Shandong Province due to the maintenance of underground equipment. It resulted in 6 deaths. And 4 persons successfully lifted the well. This accident occurred due to illegal hot work of electric welding. This mine also existed some problems such as illegal outsourcing and outsourcing instead of management. On March 15, a fire broke out in the petrochemical branch of Maoming in Guangdong. It resulted in 1 deaths and 2 injuries. On April 9, a coal and gas outburst accident occurred in Dongfeng Coal Mine in Guizhou. It resulted in 8 deaths. On May 26, an explosion occurred in Heilongjiang Huasheng Economic and Trade Co., Ltd. It resulted in 8 deaths. The accident was caused by the illegal manufacture and storage of explosives. On June 13, a gas explosion accident occurred in a market in Hubei. It resulted in 25 deaths. On July 20, an aluminum alloy company in Henan Province exploded. As the water level of the nearby Yinghe River rose sharply and exceeded the warning level This led to wall collapse and water inflow in the plant. Finally, the high-temperature solution in the alloy tank exploded. On July 24, a fire broke out in a logistics warehouse in Jilin causing 15 deaths and 25 injuries. On October 18, when maintenance personnel of Hebei Handan CR Gas Co., Ltd. were replacing valves in the pipeline well, natural gas leakage occurred. It resulted in 3 deaths. On June 16, 2022, an explosion accident occurred in the sludge drying workshop of Gansu Binnong Technology Co., Ltd. in Lanzhou. It resulted in 6 deaths and 8 injuries. On June 21, an explosion accident occurred in a doorway house in Tai'an City due to the leakage of liquefied gas tank. It resulted in 3 deaths and 10 injuries. On the same day, a gas deflagration accident occurred near the intersection, resulting in 23 injuries. There are countless safety accidents every year. The cost of accidents is painful, and the lessons of accidents are even more profound. Every accident has sounded an alarm bell for the production safety of enterprises. We must learn from the experience and reflect deeply to avoid a repeat of the tragedy.

In China, there are five hundred thousand factories and mines with varying degrees of occupational hazards. More than twenty-five million workers actually exposed to occupational hazards. China ranks first in the world in terms of the cumulative number of occupational diseases, the number of deaths and new patients. Every year, the direct loss from industrial accidents is billions of RMB, and the loss from occupational diseases is nearly 10 billion RMB. The annual economic loss caused by industrial accidents is 80 billion RMB. Thousands of families suffer from devastating disasters and incurable injuries. So following the pace of the times and the progress of science and technology, updating and optimizing safety evaluation model and method are very important.

6.1 Job risk analysis(*LEC*)

It is a simple and feasible semi-quantitative method to evaluate the risk of operators in potentially hazardous environments which was proposed by American K. J. Graham and G. F. Ginney, so it is also known as the Graham-Ginney method. And it is used in aviation and petrochemical system, but not widely. Man-machine-environment three rings theory is shown in Figure 6-1.

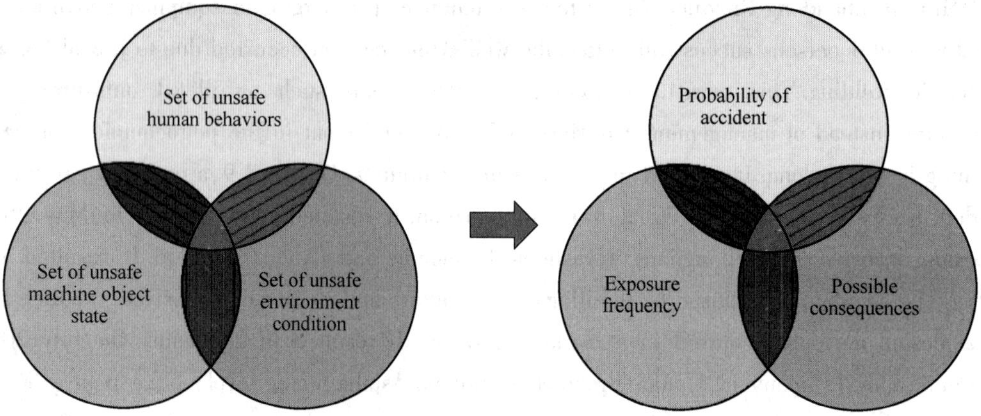

Figure 6-1 Man-machine-environment three rings theory

We should consider three factors:

(1) The possibility of accident, L;

(2) The frequency of which people are exposed to hazardous environments, E;

(3) The possible consequences of accident, C.

Taking the risk of operating conditions as the dependent variable (D), the probability of occurrence of accidents or dangerous events (L), the frequency of exposure of hazardous environments (E) and the severity of hazards (C) as independent variables, the functional equation between them is determined.

$$D = LEC \tag{6-1}$$

Therefore, the method is generally referred to as the *LEC* method.

Adopt semi-quantitative method to simplify the evaluation process, that is, different scores are given for different grades of three kinds of factors, and the product value of the scores is taken as the basis to evaluate the risk of operation conditions.

Based on the comparison of operating conditions, the evaluation team is composed of personnel familiar with operating conditions.

The evaluation team gave scores to L, E and C, and took the mean value.

The average of the three factors is multiplied to get the *D* value.

Notes: The subjectivity of experts affects the accuracy of the evaluation results, so the members of the expert group should be fully representative and keep their objectivity and independence.

(1) The possibility of accident(*L*).

The score of extremely small accident probability is set as 0.1, while the score of inevitable accident is set as10, and the score of different accident probability is determined by intermediate interpolation, see Table 6-1.

Table 6-1 The possibility of accident(*L*)

Score	The possibility of accident	Score	The possibility of accident
10	It can completely predictable	0.5	Very unlikely
6	Quite likely	0.2	Highly unlikely
3	Possible, but not often	0.1	Impossible
1	Unlikely, totally unexpected		

(2) The frequency with which people are exposed to hazardous environments(*E*).

A frequency score of 10 for exposures occurring continuously in potentially hazardous environments, 1 for exposures occurring only a few times a year and determined by intermediate interpolation, see Table 6-2.

Table 6-2 The frequency with which people are exposed to hazardous environments(*E*)

Score	Frequency of personnel exposure	Score	Frequency of personnel exposure
10	Continuous exposure	2	Monthly exposure
6	Exposure during working hours each day	1	Several exposures per year
3	Once a week, or occasionally	0.5	Very rare exposure

(3) The possible consequences of accident(*C*).

Specifies the possible outcomes of minor injuries that will require medical attention, on a score of 1; The possible outcome of causing many deaths is set at 100, with the other intermediate interpolation determining, see Table 6-3.

Table 6-3 The possible consequences of accident(*C*)

Score	Possible consequences	Score	Possible consequences
100	Great disaster. Many people died	7	Serious, serious injury
40	Disaster, several deaths	3	Major, disabling
15	Very serious. One person died	1	Need medical attention

(4) Criteria for classification of risk levels(*D*).

Experience shows that the environment with a risk score below 20 is low risk and generally acceptable. A hazard score above 320 indicates extreme danger and the operation should be stopped, see Table 6-4.

Table 6-4 Hazard score(D)

Score	Possible consequences	Score	Possible consequences
>320	It's too dangerous to continue	20-70	It can be dangerous and requires attention
160-320	It is highly dangerous and requires immediate corrective action	>20	Little danger might be acceptable
70-160	Significant risk and need to be corrected		

The evaluation can be carried out by two methods: mathematical calculation and drawing evaluation norm diagram.

The evaluation norm diagram can be drawn according to the above table and formula, see Figure 6-2.

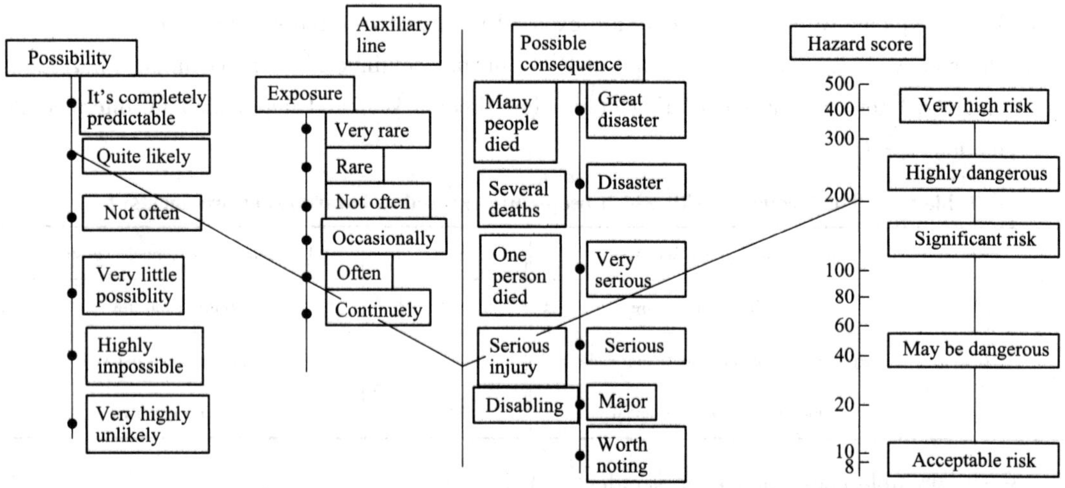

Figure 6-2 The evaluation norm diagram

The four graduated vertical lines in the figure represent the danger and its three main influencing factors. The points and the corresponding conditions are presented on these vertical lines in proportion.

Applicability: Local evaluation of operating conditions is not universally applicable. Determine quantitative criteria before the evaluation begins. For example, "perfectly predictable" is how often it happens on average, "quite likely" is how often, and so on. In this way, the risk level of each subsystem in the system can be evaluated according to the uniform standard.

The advantages and disadvantages: (1) It is simple and maneuverable, which is conducive to mastering the dangerous situation of the internal dangerous points and promoting the implementation of rectification measures. (2) Among the three factors, the possibility of accidents only has a qualitative concept, and there is no quantitative standard. During the implementation of the evaluation, the value may vary from person to person, which affects the accuracy of the evaluation results.

6. 2 Main job evaluation methods

6. 2. 1 Classification of occupational exposure to toxic substances

According to the Classification for Hazards of Occupational Exposure to Tòxicant (GB 5044—85), harm degree is divided into Level Ⅰ (extreme harm), Level Ⅱ (harm), Level Ⅲ (moderate damage), Level Ⅳ (mild harm).

6. 2. 2 Method of grading and evaluating the hazard degree of toxic operation

Classified standard of operation with exposure to toxic chemicals in the workplace (GB 12331—90).

Multiply poison damage level power coefficient (D), poisonous exercise right of labor time coefficient (L) andchroma of poison ratio (B) to get grading index (C), the poisonous exercise is divided into a Level 0 (security), Level 1 (mild harm homework), Level 2 (moderate harm homework), Level 3 (highly endangering homework), Level 4 (extreme harm homework).

6. 2. 3 Method of hazard level classification of productive dust operations

It is also called classification method for hazard degree of dust operation.

A method for evaluating the hazard level of workers exposed to productive dust operations by applying the Classification of hazard levels due to exposure to industrial (GB 5817—86) (except for radioactive dust and hazardous dust causing chemical poisoning). Table 6-5 gives the Huzardous degree scale for productive dust operation.

Select representative stations; According to the content of free SiO_2 in productive dust, the total lung flux during dust exposure time of workers and the multiple of excess dust concentration of productive dust, it shall be determined according to the table. Divided into a Level 0 (security), Level Ⅰ (mild harm homework), Level Ⅱ (moderate harm homework), Level Ⅲ (highly endangering homework), Level Ⅳ (extreme harm homework).

Table 6-5 Hazardous degree scale for productive dust operation

Free silica content in dust production	Total lung flux during exposure to dust	Production dust concentration exceeds the standard multiple							
		0	-1	-2	-4	-8	-16	-32	-64
≤10%	-4000								
	-6000								
	>6000	0	Ⅰ		Ⅱ		Ⅲ		Ⅳ

Continued Table 6-5

Free silica content in dust production	Total lung flux during exposure to dust	Production dust concentration exceeds the standard multiple							
		0	-1	-2	-4	-8	-16	-32	-64
10%-40%	-4000								
	-6000								
	>6000								
40%-70%	-4000								
	-6000								
	>6000								
>70%	-4000								
	-6000								
	>6000								

6.2.4 Method of hazard degree classification of noise operation

The hazard degree of noise in production workshops and workplaces of industrial enterprises shall be carried out by applying the method stipulated in the standard of Noise Work Classification(LD 80-95). Index classification method, divided into a Level 0 (security), Level Ⅰ (mild harm homework), Level Ⅱ (moderate harm homework), Level Ⅲ (highly endangering homework), Level Ⅳ (extreme harm homework), see Table 6-6.

Table 6-6 Noise control standards for various locations in the factory area of factories and enterprises

Category	Area	Noise limit/dBA	Remark
Ⅰ	Production workshops and workplaces	90	1. The noise limits listed in this table shall be measured and determined according to the current national standards; 2. When workers are exposed to noise for less than 8 hours a day, the noise limit can be determined according to the actual exposure time and the principle of halving the exposure time and increasing the noise limit by 3dB; 3. The indoor background noise level listed in this table refers to the indoor average noise level transmitted into the room from the outside through walls, doors and windows (the opening and closing state of doors and windows is normal) under the condition of indoor silent source
Ⅱ	The duty room, observation room and rest room set in the high-noise workshop	75	
Ⅲ	Precision assembly line, precision machining workshop location, computer room	70	
Ⅳ	The workshop belongs to the office, laboratory and design room	70	
Ⅴ	Factory office, conference room, design room, central laboratory	60	
Ⅵ	Infirmary, classroom, nursing room, nursery, worker's duty room	55	

The noise hazards in non-production workplaces should be evaluated according to the standard values stipulated in thecode for design of noise control in industrial enterprises(GBJ 87—1985).

6.2.5 The hazard degree grading evaluation method for high or low temperature operation

The hazard degree grading evaluation method for high temperature operation is a comprehensive evaluation method which applies the classification of works in hot environment(GB 4200—97) to grade the hazard degree of high temperature operation and organically combines with the Limit value of allowable continuous heat exposure time in hot environment(GB 935—89). It adopts wet bulb black bulb temperature (WBGT) (℃) and contact time of high-temperature operation to divide the high-temperature operation into four levels. The higher the level is, the greater the thermal intensity is. The operation when the wet ball black ball temperature WBGT≥25℃ is called high temperature operation.

The hazard degree grading evaluation method of low temperature operation is a method to evaluate the hazard degree of low temperature operation by applying the classification of works in cold environment(GB/T 14440—92). According to the job site of low temperature rate, operation time and temperature damage degree can be divided into levels Ⅰ, Ⅱ, Ⅲ, Ⅳ, the higher the level, the greater the cold strength. Where the average temperature of the workplace is less than 5℃.

6.2.6 Safety evaluation methods of chemical enterprises

(1) Six-stage safety evaluation method for chemical enterprises in Japan Ministry of Labor;

(2) Methods for quantitative evaluation of chemical process risk in the former Soviet Union;

(3) Chemical plant hazard rating;

(4) Safety evaluation method of thermal power station;

(5) China's metallurgy, smelting plant danger degree classification method;

(6) Safety evaluation method of machinery plant;

(7) Classification method of factory hazard;

(8) Three-stage evaluation procedure of "Safety Checklist-Risk Index Evaluation-System Safety Analysis".

6.3 Explosion accident and consequence evaluation

6.3.1 Overview

Explosion is a very sharp physical or chemical change of matter, and it is also a phenomenon in which a large amount of energy is rapidly released or rapidly converted into mechanical work in a short time. It is usually done by means of expansion of the gas.

In terms of the manifestation of material motion, explosion is a manifestation of material violent motion. A rapid increase in the motion of matter, a rapid change from one state toanother, and the release of a great deal of energy in an instant.

(1) Generally speaking, the explosion phenomenon has the following characteristics:

1) The explosion went off very quickly;

2) The pressure around the point of explosion increases sharply, creating a shock wave;

3) Make a loud or small noise;

4) The surrounding medium is shaken or the adjacent material is damaged.

(2) The explosion process is divided into two stages:

The first stage: the material energy in a certain form (constant volume, adiabatic) into a strong compression energy;

The second stage: the strong compression energy can do work to the external adiabatic expansion rapidly, causing the deformation, movement and destruction of the acting medium.

(3) It can be divided into physical explosion and chemical explosion according to the nature of explosion.

The physical explosion is a phenomenon in which the parameters of a material state (temperature, pressure, volume) change rapidly, giving off a large amount of energy and doing work in an instant. Such as boiler explosion, liquefied gas tank overpressure explosion and so on.

The chemical explosion is the rapid transformation of a substance from one chemical structure to another, giving off a great deal of energy and doing work in an instant. Such as combustible gas explosion, dust explosion and so on.

(4) Chemical explosions have two characteristics:

The chemical properties of the intermediates changed during the explosion.

The energy that forms an explosion comes from the energy released when a material changes in chemistry rapidly.

(5) There are three elements of chemical explosion:

1) Exothermic property of the reaction;

2) The rapidity of the reaction;

3) Producing gas.

A large amount of chemical energy will be released in the chemical explosion, and the explosion has a large influence range. Physical explosions release only mechanical energy, which has a small influence.

6.3.2 The energy of physical explosion

When physical explosion is caused by the rupture of a pressure vessel, the energy released by gas expansion (i. e., bursting energy) is not only related to the gas pressure and the volume of the vessel, but also to the physical phase state of the medium in the vessel. Different evaluation models are used when gas, liquid and two-phase explosions occur.

6.3.2.1 Compressed gas and water vapor container blast energy

When the medium in the pressure vessel is compressed gas, that is, it exists in gaseous form and physical explosion occurs, the explosive energy released is:

$$E_g = \frac{pV}{k-1}\left[1 - \left(\frac{0.1013}{p}\right)^{\frac{k-1}{k}}\right] \times 10^3 \qquad (6\text{-}2)$$

where E_g——The explosive energy of a gas, kJ;

p——The absolute pressure of the gas in the container, MPa;

V——Volume of container, m^3;

k——Adiabatic index of gas, that is, the ratio of specific heat at constant pressure to specific heat at constant volume.

Adiabatic index of common gas is shown in Table 6-7.

Table 6-7 Adiabatic index of common gas

Gas	Air	Nitrogen	Oxygen	Hydrogen	Methane	Ethane	Ethylene	Propane	Carbon monoxide
k	1.4	1.4	1.397	1.412	1.316	1.18	1.22	1.33	1.395
Gas	Carbon dioxide	Carbon monoxide	Nitrogen dioxide	Ammonia	Chlorine	Superheated steam	Dry saturated steam		Hydrocyanic acid
k	1.295	1.4	1.31	1.32	1.35	1.3	1.135		1.31

It can be seen from the table that the adiabatic index of air, nitrogen, oxygen, hydrogen, nitric oxide, carbon monoxide and other gases is 1.4 or approximately 1.4. If $k = 1.4$ is substituted into the above equation, then:

$$E_g = 2.5pV\left[1 - \left(\frac{0.1013}{p}\right)^{0.2857}\right] \times 10^3 \tag{6-3}$$

$$C_g = 2.5p\left[1 - \left(\frac{0.1013}{p}\right)^{0.2857}\right] \times 10^3 \tag{6-4}$$

$$E_g = C_g V \tag{6-5}$$

where C_g——Commonly used compressed gas blasting energy coefficient.

The bursting energy coefficient of compressed gas is a function of pressure. The bursting energy coefficient of compressed gas under various common pressures is shown in Table 6-8.

Table 6-8 The bursting energy coefficient of compressed gas

Gauge pressure p(MPa)	0.2	0.4	0.6	0.8	1.0	1.6	2.5
Explosion energy coefficient C_g(kJ/m³)	2×10^2	4.6×10^2	7.5×10^2	1.1×10^3	1.4×10^3	2.4×10^3	3.9×10^3
Gauge pressure p(MPa)	4.0	5.0	6.4	15.0	32.0	40.0	
Explosion energy coefficient C_g(kJ/m³)	6.7×10^3	8.6×10^3	1.1×10^4	2.7×10^4	6.5×10^4	8.2×10^4	

If $k = 1.135$ is substituted into equation (6-2), the blasting energy of dry saturated steam vessel can be obtained as:

$$E_g = 7.4pV\left[1 - \left(\frac{0.1013}{p}\right)^{0.1189}\right] \times 10^3 \qquad (6\text{-}6)$$

The blasting energy of a dry saturated steam vessel under common pressure can be calculated by the following formula:

$$E_g = C_s V \qquad (6\text{-}7)$$

where E_g——The blasting energy of steam, kJ;

V——Volume of water vapor, m^3;

C_s——Blast energy coefficient of dry saturated steam, kJ/m^3.

The bursting energy coefficient of compressed gas is shorn in Table 6-9.

Table 6-9 The bursting energy coefficient of compressed gas

Gauge pressure p(MPa)	0.3	0.5	0.8	1.3	2.5	3.0
Explosion energy coefficient C_s(kJ/m^3)	4.37×10^2	8.31×10^2	1.5×10^3	2.75×10^3	6.24×10^3	7.77×10^3

6.3.2.2 The blasting energy when the medium is all liquid

The work usually done by pressurizing a liquid is used as the energy released when a liquid pressure vessel explodes at room temperature. The formula is as follows:

$$E_L = \frac{(p-1)^2 V\beta_T}{2} \qquad (6\text{-}8)$$

where E_L——The energy released when a normal temperature liquid pressure vessel explodes, kJ;

p——Liquid pressure, Pa^{-1};

V——Volume of container, m^3;

β_T——Compressibility of liquid at pressure p and temperature T, Pa^{-1}.

For example, the compression coefficient of liquid chlorine is 2.705×10^{-3} at 50℃ and 3.175×10^{-3} at 60℃.

6.3.2.3 Blast energy of liquefied gas and high temperature saturated water

Liquefied gas and high temperature saturated water generally exist in two states of gas-liquid in the container. When the container bursts and explodes, in addition to the rapid expansion of the gas, there is also the intense evaporation process of the superheated liquid. In most cases, the saturated liquid in such a container accounts for most of the weight of the medium, and its blasting energy is much greater than that of the saturated gas.

Tip: A liquid in a saturated state is called a saturated liquid, and its corresponding steam is saturated steam, but it is only wet saturated steam at first, until the water in the steam has completely evaporated before the dry saturated steam. The temperature of steam does not increase during the process from unsaturated to wet saturation and then to dry saturation. After dry saturation, the temperature will rise and become superheated steam.

The amount of explosive energy released by a liquid when it ruptures in an overheated state:

$$E = [(H_1 - H_2) - (S_1 - S_2)T_1]W \tag{6-9}$$

where E——The blasting energy of a liquid in an overheated state, kJ;

H_1——Enthalpy of saturated liquid before explosion, kJ/kg;

H_2——Enthalpy of saturated liquid at atmospheric pressure, kJ/kg;

S_1——Entropy of saturated liquid before explosion, kJ/(kg · ℃);

S_2——The entropy of saturated liquid at atmospheric pressure, kJ/(kg · ℃);

T_1——The boiling point of a medium at atmospheric pressure, ℃;

W——The mass of the saturated liquid, kg.

The blasting energy of a saturated water container is calculated as follows:

$$E_w = C_w V \tag{6-10}$$

where E_w——The blasting energy of saturated water vessel, kJ;

V——The volume of water in container, m^3;

C_w——Saturation water blasting energy coefficient, kJ/m^3.

The bursting energy coefficient of compressed gas is shown is Table 6-10.

Table 6-10 The bursting energy coefficient of compressed gas

Gauge pressure p(MPa)	0.3	0.5	0.8	1.3	2.5	3.0
The energy coefficient C_w(kJ/m^3)	2.38×10^4	3.25×10^4	4.56×10^4	6.35×10^4	9.56×10^4	1.06×10^5

Tip: In such calculations, the work done by the expansion of the gas is generally not taken into account.

6.3.3 Shock wave and injury-destructive effect

There are three forms of blasting energy in the outward release of pressure vessel blasting:

(1) Shock wave energy;

(2) Pieces of energy;

(3) Residual deformation energy of the container.

Tip: The debris energy and the residual deformation energy of the container only account for 3%—15% of the total blasting energy.

6.3.3.1 Shock wave

The shock wave is formed by the superposition of compression waves. It is the compression wave propagated by the wave front in the form of inrush in the medium. When the vessel is broken, a large number of high-pressure gases inside the vessel will rush out and disturb the surrounding air by the shock wave, making its state (pressure, temperature, density, etc.) undergo a sudden change. Its propagation velocity is greater than the sound velocity of the disturbed medium, and the propagation of such disturbance in the air becomes the shock wave.

At a certain distance from the detonation center, the maximum positive pressure propagated to this point after the beginning of the explosion is called the overpressure Δp on the front of the impact wave.

Shock wave damage, damage effect criteria are: overpressure criteria, impulse criteria, overpressure-impulse criteria, etc.

The damage of overpressure wave to human body and the damage to buildings are shown in Table 6-11 and Table 6-12.

Table 6-11 Damage effect of shock wave overpressure on human body

The overpressure Δp(MPa)	Damage
0.02—0.03	Slight damage
0.03—0.05	Auditory organ damage or fracture
0.05—0.10	Severe internal damage or death
>0.10	Most of the deaths

Table 6-12 Damage effect of shock wave overpressure on buildings

The overpressure Δp(MPa)	Damage	The overpressure Δp(MPa)	Damage
0.005—0.006	Doors and windows are partially broken	0.06—0.07	The pillar of the wooden building is broken and the frame is loose
0.006—0.015	Most of the pressurized doors and window glass is broken	0.07—0.10	Brick wall collapsed
0.015—0.02	Window frame damage	0.10—0.20	Shockproof reinforced concrete was destroyed and small houses collapsed
0.02—0.03	Wall cracks		
0.04—0.05	Big cracks appeared in the walls and the tiles fell off the house	0.20—0.30	Failure of large steel frame structure

6.3.3.2 The overpressure of the shock wave

The overpressure in the front of the shock wave is related to the energy that generates the shock wave as well as the distance from the center of the explosion. The relationship between the overpressure of the shock wave and the distance of the explosion center is:

$$\Delta p \propto R^{-n} \tag{6-11}$$

where Δp——Impact wave front overpressure, MPa;

R——Distance from the center of the explosion, m;

n——Attenuation coefficient.

The attenuation coefficient varies with the overpressure in the air and is 2.5—3.0 in the vicinity of the explosion center. When the overpressure is within a few atmospheres, it is 2; When the overpressure is less than 1atm, the attenuation coefficient is 1.5.

The experimental data show that when different quantities of similar explosives explode, if the ratio of R to the distance from the explosion center is equal to the ratio of the explosive quantity Q to the cubic root of the explosive, the shock wave overpressure generated will be the same. The formula is as follows:

$$\frac{R}{R_0} = \sqrt[3]{\frac{q}{q_0}} = \alpha \rightarrow \Delta p = \Delta p_0 \tag{6-12}$$

where R——Distance between target and explosion center, m;

R_0——A considerable distance from the target to the center of the reference explosion, m;

q_0——Reference explosion energy, TNT, kg;

q——The amount of energy expended to produce a shock wave during an explosion, TNT, kg;

Δp——The overpressure at the target, MPa;

α——Simulation ratio of explosive detonation test.

Can also be written as:

$$\Delta p(R) = \Delta p_0\left(\frac{R}{\alpha}\right) \tag{6-13}$$

The above formula can be used to determine the overpressure at various distances when an arbitrary dose explodes, based on the overpressure measured by tests of some known dose.

To calculate the damage and damage effect of the pressure vessel explosion on the target, the following procedure can be followed.

(1) First, according to the properties of the medium in the container, the blasting energy E is calculated;

(2) The blasting energy q is converted into TNT equivalent q_0, because the explosion energy released by 1kg TNT explosion is 4230—4836kJ/kg, and the average value is 4500kJ/kg, so the relation is

$$q = \frac{E}{q_{\mathrm{TNT}}} = \frac{E}{4500} \tag{6-14}$$

(3) Calculate the simulation ratio of the explosion:

$$\alpha = \left(\frac{q}{q_0}\right)^{\frac{1}{3}} = \left(\frac{q}{1000}\right)^{\frac{1}{3}} = 0.1q^{\frac{1}{3}} \tag{6-15}$$

(4) Find the corresponding distance in 1000kg TNT explosion test:

$$R_0 = \frac{R}{\alpha} \tag{6-16}$$

(5) According to the value in the table to find the distance to R_0 of the overpressure Δp_0 (intermediate value by insertion method), this is the distance to R of the overpressure;

(6) According to the Δp value of overpressure, find out the damage to people and buildings from the table.

6.3.3.3 Steam cloud explosion shock wave damage, damage radius

Explosive gases are stored in liquid form, and vapor cloud explosions may occur if a transient leak is followed by delayed ignition or gaseous storage is leaked into the air when meeting a fire source. Generally speaking, only compression energy and heat energy can form vapor clouds alone.

According to the recommendations of the Netherlands Organisation for Applied Scientific Research (TNO), the damage radius of the steam cloud explosion shock wave can be predicted by pressing the following formula:

$$R = C_s (NE)^3 \tag{6-17}$$

where R——The damage radius, m;

E——The explosion energy, kJ, $E = V \cdot H_c$; V is the volume of combustible gas involved in the reaction, m^3; H_c is the high calorific value of combustible gas, kJ/m^3, the value is shown in Table 6-13;

N——The efficiency factor, whose value is related to the ratio of loss caused by continuous expansion of combustion concentration and the amount of mechanical energy obtained from fuel combustion, is generally set as $N = 10\%$;

C_s——The empirical constant, depending on the damage level, is shown in Table 6-14.

Table 6-13 The high calorific value of some gases

Name		High calorific value	Name	High calorific value
Hydrogen		12770	Ethylene	6419
Ammonia		17250	Acetylene	58985
Benzene		47843	Propane	101828
Carbon monoxide		17250	Propylene	94375
Hydrogen sulfide	Creat SO_2	25708	n-butane	134026
	Creat SO_3	30146	Isobutane	132016
Methane		39860	Butane	121883
Ethane		70425		

Table 6-14 Damage level table

Damage level	Value	Equipment damage	Human injury
1	0.03	Serious damage to buildings and processing equipment	1% died from lung injuries; >50% the eardrum rupture; >50% injured by debris

Continued Table 6-14

Damage level	Value	Equipment damage	Human injury
2	0.06	Damage to building appearance, which can't be repaired	1% the eardrum rupture; 1% injured by debris
3	0.15	Glass broken	Injured by debris
4	0.4	10% of the glass broken	

6.3.4 Various VCE computing models

6.3.4.1 Introduction to TNT equivalent method

The basic idea of this method is to convert the vapor cloud mass into an equivalent amount of TNT. In general, only a small part of the heat from fuel combustion is expressed in the form of shock waves, so this method assumes that only a certain percentage (1 to 10 percent) of the fuel contributes to overpressure formation. Its basic formula is

$$W_{TNT} = \alpha_e \frac{W_f H_f}{H_{TNT}} \tag{6-18}$$

where W_f——The mass of vapor cloud, kg;

H_f——Combustion heat of fuel, MJ/kg;

H_{TNT}——The detonation energy of TNT, 4.52 MJ/kg;

α_e——The efficiency coefficient.

Figure 6-3 and Figure 6-4 give the explosive overpressure diagram and explosive impulse chart of ground TNT explosions, respectively.

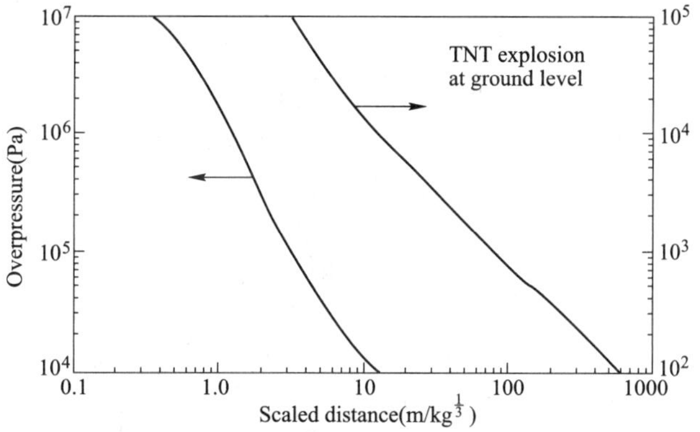

Figure 6-3 Explosive overpressure diagram of ground TNT explosions

6.3.4.2 Introduction to the TNO method

This method describes the explosion problem of hemispherical gas clouds located on the ground, as shown in the Figure 6-5. The basic idea is to calculate the combustion energy of the gas cloud and calculate the explosion overpressure at a certain distance by the characteristic explosion

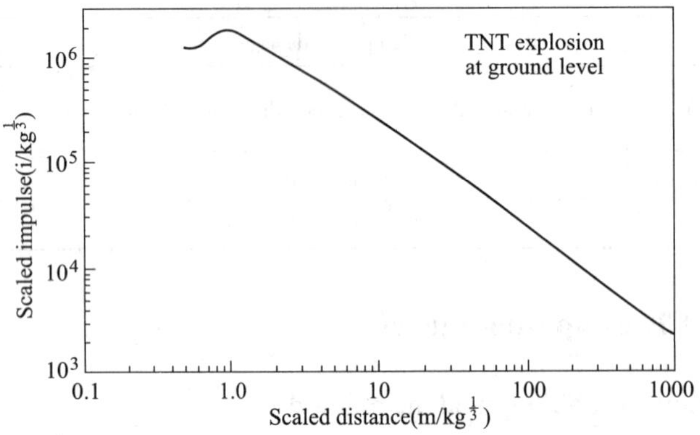

Figure 6-4 Explosive impulse chart for ground TNT explosions

length. Definition of characteristic explosion length:

$$L_0 = \left(\frac{2}{3} \times \Pi \times r_0^3 \times \frac{E_c}{p_0}\right)^{\frac{1}{3}} \tag{6-19}$$

where L_0——The characteristic explosion length, m;

r_0——The radius of a hemispherical cloud, m;

E_c——Gas clouds burn energy density, J/m^3;

p_0——Environmental pressure, Pa.

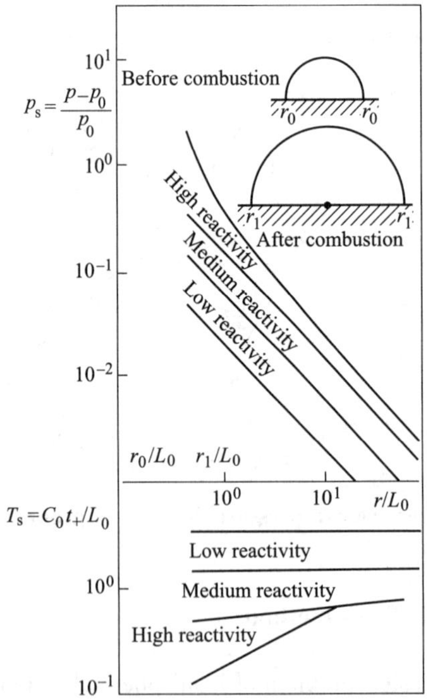

Figure 6-5 Graph of TNO method calcuate

6. 3. 4. 3 Introduction to the Pluripotent method

Compared with the previous two methods, the multi-energy method emphasizes the influence of obstacles and boundary constraints on the consequences of explosion. The method divides the vapor cloud into several volumes according to the different degree of gas constraint, and each volume corresponds to an explosion intensity index(1-10). An explosion intensity index of 1 corresponds to an unrestrained or accessible vapor cloud explosion, while 10 corresponds to a vapor cloud detonation. The velationship between time and distance of static overpressure is shown in Figure 6-6. Static overpressure and dynamic overpressure are shown in Figure 6-7 and Figure 6-8, respectively.

$$R_0 = R\left(\frac{p_0}{E}\right)^{\frac{1}{3}} \tag{6-20}$$

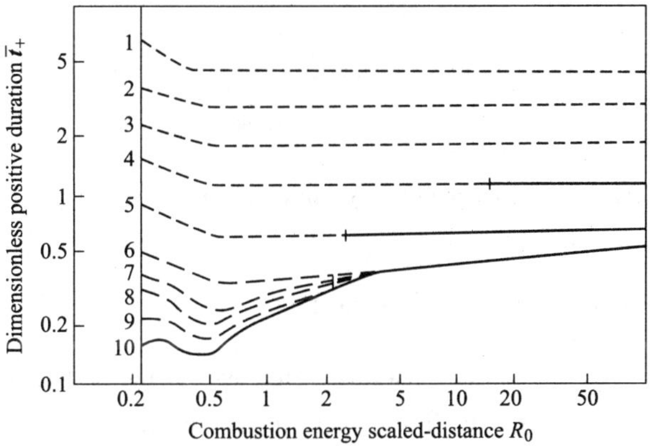

Figure 6-6 The relationship between time and distance of static overpressure

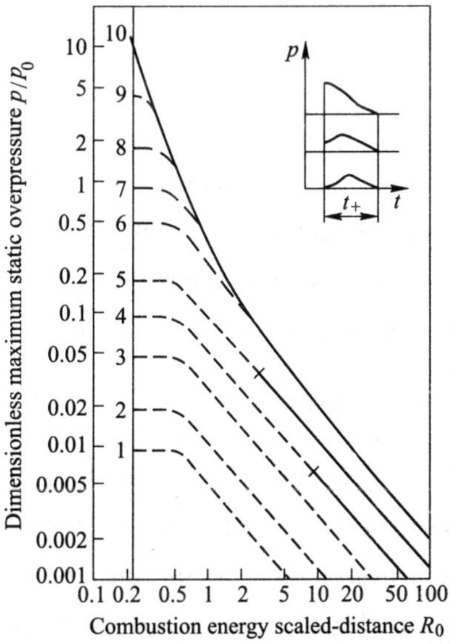

Figure 6-7 Static overpressure

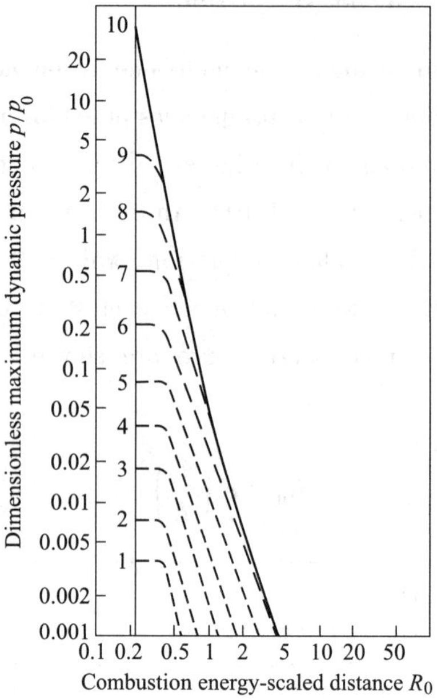

Figure 6-8 Dynamic overpressure

6.3.4.4 Introduction to the British Gas method

This method is used to calculate the explosion consequences of non-detonation gas clouds (methane with less than 5% ethane) and is actually an improvement on the TNT equivalent method. In this method, 400kPa overpressure will be generated in the identified explosion area, the efficiency coefficient of TNT equivalent method increases to 0.2, and the energy proportional coefficient is 10.

6.3.4.5 Introduction to Shell methods

This method argues that the previous methods underestimate the effect of explosions in the far field. The basic principle of this method is the same as that of the multi-energy method, except that this method estimates the explosion overpressure through an inverse proportional attenuation law. In addition, this method is associated with a decision tree, as shown in Figure 6-9.

6.3.4.6 Comparative analysis of methods

Comparative analysis of methods is shown in Figure 6-10. These empirical prediction models differ greatly in the near field but tend to be consistent in the far field. Since TNT equivalent method is easy to overestimate the explosion overpressure in the near field, TNO method has the defect of ignoring the influencing factors of obstacles, so in practical application, multi-energy method, British Gas method and Shell method should be given priority.

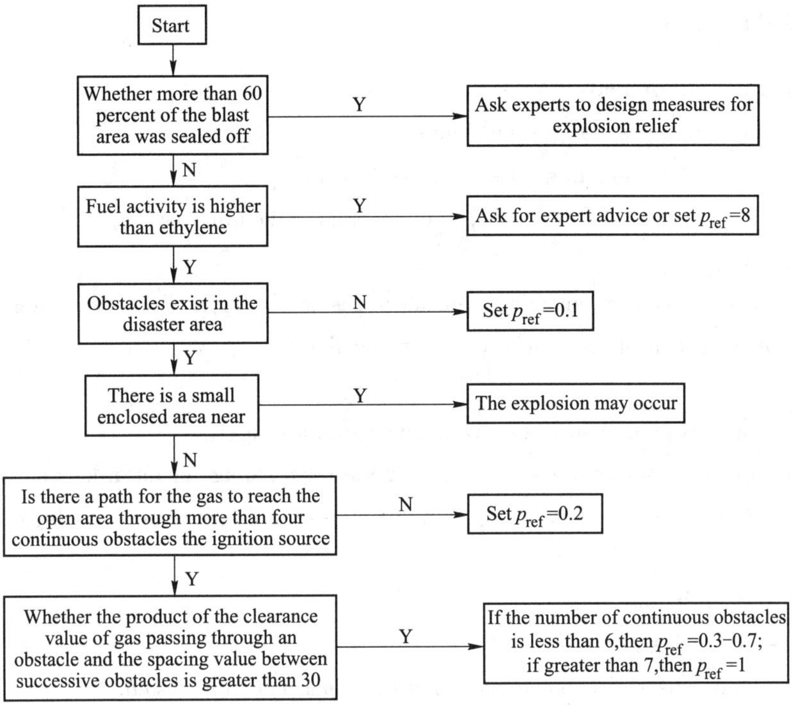

Figure 6-9 Decision tree

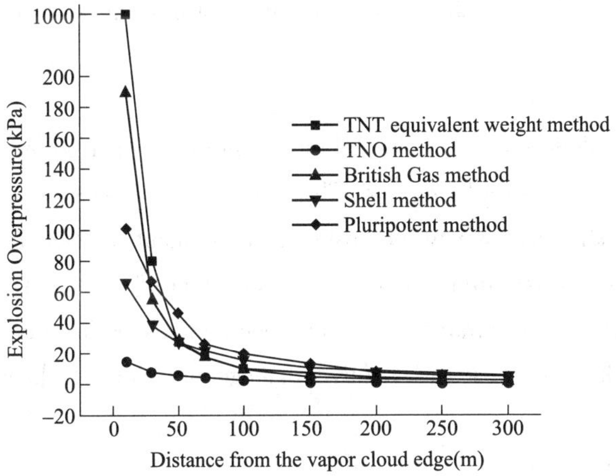

Figure 6-10 Comparative analysis of methods

6.4 Scope(or damage) evaluation method

6.4.1 Leak

Due to equipment damage or operation errors caused by leakage, a large number of flammable, explosive, toxic and harmful substances released, will lead to fire, explosion, poisoning and other major accidents. So the consequence analysis starts with the leakage analysis.

6.4.1.1 Leakage analysis

A Major leaking equipments

(1) Pipes: including pipes, flanges and joints;

(2) Flexible connector: including hoses bellows and hingers;

(3) Filter: it is composed of filter body, pipeline, filter screen and so on;

(4) Valve;

(5) Pressure vessels, reactors: including separators, gas scrubbers, reactors, heat exchangers, various tanks and containers commonly used in chemical production;

(6) Pump;

(7) Compressor: including centrifugal, axial and reciprocating;

(8) Storage tank: a vessel or pressure vessel for the open storage of hazardous substances;

(9) Pressurize or freeze gas containers: including storage tanks, pressure vessels, or tanker trucks for open burial;

(10) Torch burner or flare tube.

B The cause of the leak

There are four main causes of leakage accidents from man-machine system.

a Design mistakes

(1) Basic design error; (2) Improper material selection; (3) Improper arrangement; (4) The machine is not suitable; (5) The selection of measuring instruments isn't appropriate; (6) No liquid level gauge of tank.

b Design reasons

(1) The processing does not meet the requirements; (2) Poor processing quality; (3) Low accuracy of construction and installation; (4) The quality of the selected standard products is not up to standard; (5) The acceptance of installed equipment is not in accordance with the "Mechanical equipment installation engineering Acceptance"; (6) After long-term use, the equipment is not repaired in accordance with the specified inspection period; (7) The measuring instrument is not checked regularly; (8) Valve damage or switch leakage; (9) Poor quality of equipment accessories.

c Administrative reasons

(1) There is no complete safe operation procedure; (2) Indifference to safety and lack of timely resolution of identified problems; (3) The supervision and inspection system is not strictly implemented; (4) Command error; (5) Put untrained people on the job; (6) The overhaul system is lax.

d Human errors

(1) Misoperation, violation of operating rules; (2) Wrong judgment; (3) To take off the post without authorization; (4) Lack of concentration; (5) It is not known what to do when an abnormal condition appears.

C Consequences of leakage

The consequences of leakage vary depending on the substance leaked:

a Flammable gas leakage

Combustion or explosion will occur when combustible gas is mixed with air after leakage and reaches combustion limit when it meets ignition source. The consequences of a leak vary depending on the time of the fire.

(1) A fire immediately: Combustible gas is ignited when it leaks out of the container and diffuses into flames, which can quickly endanger the leakage site, but rarely affect the outside of the plant.

(2) Lag of fire: After the combustible gas is discharged, it will mix with the air to form a combustible vapor cloud and drift with the wind, which will lead to explosion or detonation when it meets the fire source, causing a large range of damage.

b Toxic gas leakage

Toxic gas leakage after the formation of clouds in the air spread, the impact area is large.

c Liquid leakage

(1) Liquid leakage under normal temperature and pressure: The liquid accumulates to form the liquid pool, if encounters the fire source to be able to produce the pool fire.

(2) Pressurized liquefied gas leakage: Some liquids evaporate instantaneously when leaking, and some may evaporate completely during the process.

(3) Cryogenic liquid leakage: When this liquid leaks, it forms a liquid pool, and the evaporation capacity is lower than that of pressurized liquefied gas and higher than that of liquid under normal temperature and pressure.

6.4.1.2 Leakage calculation

A Liquid leakage model

The leakage rate of liquid can be calculated by Bernoulli equation of fluid mechanics, and the leakage rate is:

$$Q_0 = C_d A \sqrt{\frac{2(p - p_0)}{\rho} + 2gh} \tag{6-21}$$

where Q_0——Fluid leakage rate, kg/s;

C_d——Liquid leakage coefficient, select according to Table 6-15;

A——Cleft area, m^2;

ρ——Leakage rate, kg/m^3;

p——Pressure of medium in container, Pa;

p_0——Environmental stress, Pa;

g——Gravitational acceleration, 9.8 m/s^2;

h——Liquid level height above the breach, m.

Table 6-15 Liquid leakage coefficient C_d

Reynolds number(*Re*)	Crack shape		
	Circular(Polygon)	Triangle	Rectangle
>100	0. 65	0. 60	0. 55
≤100	0. 50	0. 45	0. 40

When the liquid in the container is superheated, it evaporates abruptly as the pressure decreases as it passes through the breach. In this case, the percentage F of the liquid directly evaporated at the time of leakage can be calculated as follows:

$$F = c_p \frac{T - T_0}{H} \tag{6-22}$$

where c_p——The specific heat of the liquid at constant pressure, J/(kg · K) ;

T——The temperature of the liquid before leakage, K;

T_0——The boiling point of a liquid at atmospheric pressure, K;

H——Heat of vaporization of liquid, J/kg.

According to Equation(6-21), the result is almost always between 0 and 1. As a rule of thumb, when $F>0.2$, a liquid pool is generally not formed. When $F<0.2$, there is a linear relationship between the ratio of F and the liquid taken away, when $F=0.2$, there is no liquid taken away, and when $F=0.1$, 50% of the liquid is taken away.

B Gas leakage model

The rate at which a gas escapes from a crack depends on its state of flow. Therefore, when calculating the leakage amount, we should first judge whether the gas flow at the time of leakage belongs to the sonic or subsonic flow. The former is called the critical flow, while the latter is called the subcritical flow.

When the current equation holds, the gas flow belongs to the sonic flow:

$$\frac{p_0}{p} \leqslant \left(\frac{2}{k + 1}\right)^{\frac{k}{k-1}} \tag{6-23}$$

When the current formula is established, gas flow is subsonic flow:

$$\frac{p_0}{p} \geqslant \left(\frac{2}{k + 1}\right)^{\frac{k}{k-1}} \tag{6-24}$$

where p——Pressure of medium in container, Pa;

p_0——Environmental stress, Pa;

k——Adiabatic index for a gas, the ratio of the specific heat capacity c_p at constant pressure to the specific heat capacity c_V at constant pressure. When the gas flows in a sonic manner, its leakage is Equation(6-22); when the gas flows in a subsonic manner, its leakage is Equation(6-23).

$$Q_0 = C_d A p \sqrt{\frac{Mk}{RT}\left(\frac{2}{k + 1}\right)^{\frac{k+1}{k-1}}} \tag{6-25}$$

$$Q_0 = YC_dAp\sqrt{\frac{Mk}{RT}\left(\frac{2}{k+1}\right)^{\frac{k+1}{k-1}}} \tag{6-26}$$

$$Y = \sqrt{\frac{2}{k-1}\left(\frac{k+1}{2}\right)^{\frac{k+1}{k-1}}\left(\frac{p}{p_0}\right)^{\frac{2}{k}}\left[1-\left(\frac{p_0}{p}\right)^{\frac{k-1}{k}}\right]} \tag{6-27}$$

where C_d——Gas leakage coefficient, Take 1.0 when the crack shape is circular, 0.95 when triangular, and 0.90 when rectangular;

Y——Gas expansion factor;

M——The molecular weight;

ρ——Gas density, kg/m^3;

R——Gas constant, J/(mol · K);

T——The gas temperature, K.

When the material in the container decreases with leakage or the pressure decreases and the discharge velocity is affected, the calculation of leakage velocity is accurate. If the flow velocity is small and the ignition time is short, the initial discharge velocity can be used in the consequence calculation. Otherwise, the equivalent leakage velocity should be calculated.

C Two-phase flow leakage model

When superheated liquid leakage occurs, sometimes there will be gas and liquid two-phase flow, which has the dual characteristics of gas leakage and liquid leakage.

The leakage velocity of uniform two-phase flow can be calculated by the following formula:

$$Q_0 = C_dA\sqrt{2\rho(p-p_c)} \tag{6-28}$$

where Q_0——Leakage rate of two-phase flow mixture, kg/s;

C_d——Leakage coefficient of two-phase flow mixture, can take 0.8;

A——Cleft area, m^2;

p——The pressure of the two-phase mixture, Pa;

p_c——Critical pressure, Pa, can take $p_c = 0.55$Pa;

ρ——The mean density of a two-phase mixture, kg/m^3, it is calculated by the following formula:

$$\rho = \frac{1}{\frac{F_v}{\rho_1} + \frac{1-F_v}{\rho_2}} \tag{6-29}$$

where ρ_1——The density of vapor from a liquid, kg/m^3;

ρ_2——The liquid density, kg/m^3;

F_v——The ratio of the evaporated liquid to the total liquid, which can be calculated by the following formula:

$$F_v = \frac{c_p(T-T_c)}{H} \tag{6-30}$$

where c_p——Specific heat at constant pressure for a two-phase mixture, J/(kg · K);
T——The temperature of a two-phase compound, K;
T_c——Critical temperature, K;
H——Heat of vaporization of liquid, J/kg.

When $F_v>1$, indicating that the liquid will all evaporate the gas, the matter should be calculated according to the gas leakage formula; If the F_v is small, it can be calculated approximately according to the liquid leakage formula.

6.4.2 Diffusion

6.4.2.1 Overview

The characteristics of the leaking material are varied and strongly influenced by the original conditions. For most substances that leak from a container, only the forms of gas mass release are discussed, post-leak liquid diffusion, jet diffusion, and adiabatic diffusion.

6.4.2.2 Diffusion model after leakage

A Liquid diffusion model

The liquid immediately diffuses to the surface and forms a pool. The amount of liquid in the pool will remain constant when the liquid evaporation rate is equal to the leakage rate.

a The liquid pool area

If the leaking liquid reaches the artificial boundary, then the pool area is the area enclosed by the artificial boundary. If not, the leakage point of the liquid is assumed to be a flat cylinder at the center and diffuses on a smooth plane. At this time, the radius r of the liquid pool is calculated by the following formula:

Instantaneous leakage (leakage time does not exceed 30s)

$$r = \left(\frac{8gm}{\pi p}\right)^{\frac{\sqrt{t}}{4}} \tag{6-31}$$

Continuous leakage (leakage lasts more than 10min)

$$r = \left(\frac{32gmt^3}{\pi p}\right)^{\frac{1}{4}} \tag{6-32}$$

where r——Radius of liquid pool, m;
m——The quality of the leaking liquid, kg;
g——Gravitational acceleration, 9.8m/s^2;
p——Liquid pressure in equipment, Pa;
t——Leakage time, s.

b Evaporation

The evaporation of liquid in the tank can be divided into flash evaporation, heat evaporation and mass evaporation according to its mechanism.

(1) Flash evaporation: After the superheated liquid leaks, the direct evaporation of the liquid

due to its own heat is called flash evaporation. When flash occurs, the liquid evaporation rate Q_1 can be calculated by the following formula:

$$Q_1 = F_v \cdot \frac{m}{t} \tag{6-33}$$

where F_v——The ratio of directly evaporated liquid to total liquid;
m——The quality of the leaking liquid, kg;
t——Flash evaporation time, s.

(2) Heat of evaporation: When $F_v<1$, the liquid flash evaporation is not complete, and part of the liquid forms a liquid pool on the ground and absorbs heat from the ground to vaporize, which is called heat evaporation. Heat evaporation rate Q_1 is calculated by the following formula:

$$Q_1 = \frac{KA_1(T_0 - T_b)}{H\sqrt{\pi \alpha t}} + \frac{KNuA_1}{HL}(T_0 - T_b) \tag{6-34}$$

where A_1——The liquid pool area, m^2;
T_0——The environment temperature, K;
T_b——The liquid boiling point, K;
L——The length of the liquid pool, m;
α——Thermal diffusion coefficient, m^2/s, see Table 6-16;
K——Coefficient of thermal conductivity, J/(m · K), see Table 6-16;
t——The evaporation time, s;
Nu——Nuschert number.

Table 6-16 Heat transfer properties of certain surfaces

Ground	K(J/(m · K))	α(m^2/s)
Cement	1.1	1.29×10^{-7}
Land	0.9	4.3×10^{-7}
Dry land	0.3	2.3×10^{-7}
Wetland	0.6	3.3×10^{-7}
Gravel land	2.5	11.0×10^{-7}

(3) Mass evaporation: When the ground surface heat transfer stops, the heat evaporation stops and the liquid evaporates due to the movement of air over the surface of the pool, which is called mass evaporation. The evaporation rate is:

$$Q_1 = \alpha Sh \frac{A}{L}\rho_1 \tag{6-35}$$

where α——Molecular diffusion coefficient, m^2/s;
Sh——Sherwood number;
A——The liquid pool area, m^2;
L——The length of the liquid pool, m;
ρ_1——Density of liquid, kg/m^3.

B Jet diffusion model

When gas leaks, it is ejected from the crack to form a gas jet. In the calculation of injection, the equivalent diameter of injection orifice should be used. The orifice diameter of equivalent injection is calculated by the following formula:

$$D = D_0\sqrt{\frac{\rho_0}{\rho}} \tag{6-36}$$

where D——Equivalent jet aperture, m;

D_0——Fissure aperture, m;

ρ_0——The density of the leaking gas, kg/m^3;

ρ——The density of the gas in ambient conditions, kg/m^3.

If the gas leakage can be instantaneous to the ambient temperature, pressure. That is, $\rho_0 = \rho$, then $D_0 = D$.

a The concentration distribution of the spray

The gas density $C(x)$ at the point x from the orifice on the injection axis is:

$$C(x) = \frac{\dfrac{b_1 + b_2}{b_1}}{0.32\dfrac{x}{D}\cdot\dfrac{\rho}{\sqrt{\rho_0}} + 1 - \rho} \tag{6-37}$$

$$b_1 = 50.5 + 48.2\rho - 9.95\rho^2$$

$$b_2 = 23.0 + 41.0\rho$$

The symbolic meaning is the same as before.

The gas concentration at any point x on the jet axis and perpendicular to the jet axis is:

$$\frac{C(x,y)}{C(x)} = e^{-b_2(y/x)^2} \tag{6-38}$$

where $C(x,y)$ ——Gas concentration at Y point in the plane x away from the break and perpendicular to the jet axis, kg/m^3;

$C(x)$ ——Other densities along the jet axis from the crack x, kg/m^3;

b_2——Distribution function, the meaning is the same as before;

y——Distance from the target point to the jet axis, m.

b The velocity distribution of the jet axis

The injection velocity decreases with the increase of the axis distance until the injection velocity equals the wind speed at a point along the axis, which is called the critical point. The gas cloud after the critical point no longer conforms to the injection law.

The velocity distribution along the axis of the jet is obtained from the following equation:

$$\frac{V(x)}{V_0} = \frac{\rho_0}{\rho} \cdot \frac{b_1}{4}\left(0.32\frac{x}{d} \cdot \frac{\rho}{\rho_0} + 1 - \rho\right)\left(\frac{D}{x}\right)^2 \tag{6-39}$$

where ρ_0——The density of the leaking gas, kg/m^3;

ρ——Gas density in ambient conditions, kg/m^3;

D——Equivalent jet aperture, m;

b_1——Distribution function, the meaning is the same as before;

x——The distance on the jet axis from a point in the breach, m;

$V(x)$——The velocity at the point x on the jet axis, m/s;

V_0——The initial velocity of the injection, which is equal to the velocity at which the gas escapes through the crack, m/s;

$$V_0 = \frac{Q_0}{C_d \rho \pi \left(\frac{Q_0}{2}\right)^2} \tag{6-40}$$

where Q_0——Gas leakage speed, km/s;

C_d——Gas leakage coefficient.

When the concentration at the critical point is less than the allowable concentration, it is analyzed only according to the injection. If the concentration at this point is greater than the allowable concentration, the diffusion of the leakage gas in the atmosphere needs to be further analyzed.

C Adiabatic diffusion model

A period of rapid diffusion following an instantaneous leakage of a flash liquid or pressurized gas is said to be adiabatic diffusion, assuming that the process is so rapid that there is no time for exchange between the mixed air mass and the surrounding environment.

Adiabatic diffusion is divided into two stages. In the first stage, the air mass diffuses outward to atmospheric pressure, during which the air mass gains kinetic energy, which is called "diffusion energy". In the second stage, the diffusion energy can push the air mass outward again, making the turbulent mixed air enter the air mass, thus expanding the range of the air mass.

a Air mass diffusion energy

In the first stage of air mass diffusion, part of the internal energy of the diffusing gas is used to increase kinetic energy and do work to the surrounding atmosphere. Suppose the process at this stage is reversible adiabatic and isentropic.

(1) Gas leakage diffusion energy according to the change of internal energy, the formula of diffusion energy is as follows:

$$E = c_V(T_1 - T_2) - 0.98p_0(V_2 - V_1) \tag{6-41}$$

where E——Gas diffusion energy, J;

c_V——Constant volume specific heat, J/(mol · K);

T_1——Initial air mass temperature, K;

T_2——The temperature at which air mass pressure drops to atmospheric pressure, K;

p_0——Environmental stress, Pa;

V_1——Initial volume of air mass, m^3;

V_2——The volume of a gas at atmospheric pressure, m^3.

(2) Flash evaporation liquid leakage diffusion energy. The evaporation energy of the vaporized vapor mass can be calculated as follows:

$$E = [H_1 - H_2 - T_b(S_1 - S_2)]W - 0.98(p_1 - p_0)V_1 \tag{6-42}$$

where E——Flash evaporation liquid diffusion energy, J;

H_1——Initial enthalpy of leaking liquid, J/kg;

H_2——Final enthalpy of the leaking liquid, J/kg;

T_b——The boiling point of the liquid, K;

S_1——The entropy of the liquid before evaporation, J/(kg · K);

S_2——The entropy of the liquid after evaporation, J/(kg · K);

W——Liquid evaporation, kg;

p_1——Initial pressure, Pa;

p_0——Ambient pressure, Pa;

V_1——The initial volume, m^3.

b Air mass radius and concentration

Driven by the diffusion energy, the air mass diffuses outward and mixes with the surrounding air turbulent.

(1) Inner radius and concentration: The inner radius R_1 and concentration C of the air mass are time functions, expressed as follows:

$$R_1 = 2.72\sqrt{k_d t}$$
$$C = \frac{0.0059TV_0}{\sqrt{(k_d t)^3}} \tag{6-43}$$

where t——Diffusion time, s;

V_0——Volume of gas at standard temperature and pressure, m^3;

k_d——Turbulent diffusion coefficient, calculating according to the following formula:

$$k_d = 0.0137\sqrt[3]{V_0} \cdot \sqrt{E} \cdot \left(\frac{\sqrt[3]{V_0}}{\sqrt{E}}\right)^{\frac{1}{4}} \tag{6-44}$$

As mentioned above, the second stage does not end until the central diffusion velocity drops to a certain value. The selection of critical speeds is random and unstable. Suppose the diffusion velocity at the end of diffusion is 1m/s, then the inner radius R_1 and the concentration C at the end of diffusion can be calculated according to the following formula:

$$R_1 = 0.08837E^{0.3}V_0^{\frac{1}{3}}$$
$$C = 172.95E^{-0.9} \tag{6-45}$$

(2) Outer radius and concentration: The size of the outer layer of the air mass at the end of the

second stage can be obtained by experimental observation, that is, the outer air mass radius R_2 at the end of diffusion can be obtained from the following equation:

$$R_2 = 1.456R_1 \tag{6-46}$$

where R_2, R_1——The radius of the outer and inner layers of the air mass, m.

The concentration of outer layer gas isgaussian distribution from inside to outside.

6.4.3 Fire

Flammable, explosive gases and liquids will be ignited when they meet the ignition source after leakage. They burn by pool fire, jet fire, fireball and burst fire.

6.4.3.1 Evaluation model of pool fire flame and radiation intensity

Flammable liquid(gasoline, diesel, etc.) after leaking back to the ground to form pool liquid, or flow to the water surface and cover the water surface, meet the flame burning and pool fire.

Open environment of pool fire, belongs to the non premixed combustion, liquid local oxygen and high temperature cracking occurred when burning, the carbon black precipitate, form the luminous flame, the flame of the other ingredients, such as radiation ability of CO_2, H_2O and other three atomic gas radiation ability compared with carbon black particles is low, so the main flame radiation as the radiation of carbon black particles.

A Combustion rate

When the boiling point of the flammable liquid in the liquid pool is higher than the ambient temperature, the burning rate per unit area on the liquid surface is:

$$\frac{\mathrm{d}m}{\mathrm{d}t} = \frac{0.001H_c}{C_p(T_b - T_0) + H} \tag{6-47}$$

where $\mathrm{d}m/\mathrm{d}t$——Burning rate per unit surface area, kg/(m^2 · s);

H_c——Heat of liquid combustion, J/kg;

C_p——The specific heat of the liquid at constant pressure, J/(kg · K);

T_b——The boiling point of the liquid, K;

T_0——The environment temperature, K;

H——Heat of vaporization of liquid, J/kg.

When the boiling point of a liquid is lower than the ambient temperature, such as pressurized liquid gas or frozen liquid gas, the combustion rate per unit area is:

$$\frac{\mathrm{d}m}{\mathrm{d}t} = \frac{0.001H_c}{H} \tag{6-48}$$

Calculation of pool diameter:

$$D = \left(\frac{3S}{3.14}\right)^{\frac{1}{2}} \tag{6-49}$$

Where there is a protective levee, the area of the tank shall be the area enclosed by the protective levee, such as the oil tank or the oil tank area; When there is no protective embankment, such as oil pipeline, the maximum liquid pool area S is calculated according to the amount of liquid leakage and the surface property:

$$S = \left(\frac{W}{H_{min}\rho}\right)^{\frac{1}{2}} \tag{6-50}$$

where H_{min}——Minimum reservoir thickness.

Minimum reservoir thickness of different surface is shown in Table 6-17.

Table 6-17 Minimum reservoir thickness of different surfaces

Nature of the ground	Minimum reservoir thickness H_{min}/m	Nature of the ground	Minimum reservoir thickness H_{min}/m
The grass	0. 020	Concrete ground	0. 005
Rough ground	0. 025	Calm water surface	0. 0018
Flat ground	0. 010		

B Flame height

Let the liquid pool be a round pool with a radius of R (Figure 6-11), and its flame height is calculated according to Thomas's empirical formula:

$$\frac{L}{D} = 42\left[\frac{\frac{dm}{dt}}{\rho_0(gD)^{\frac{1}{2}}}\right]^{0.61} \tag{6-51}$$

where L——Flame height, m;

D——The liquid pool diameter, m;

ρ_0——Ambient air density, kg/m^3;

g——Gravitational acceleration, 9. 8m/s^2;

dm/dt——Combustion rate, kg/(m^2 · s).

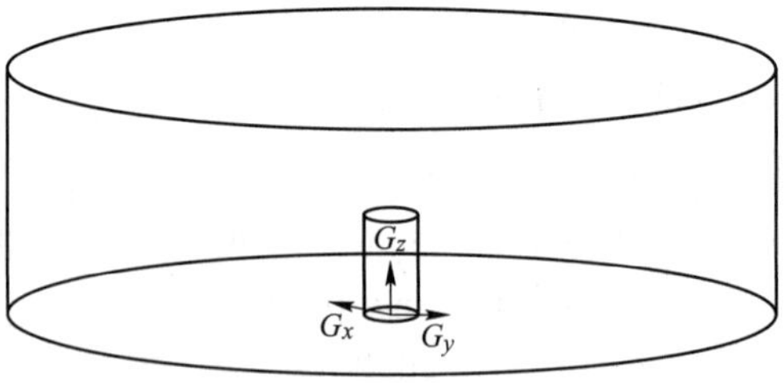

Figure 6-11 Geometric model diagram

This experiment is obtained from the woodpile experiment, which is slightly larger than the real pool fire flame value.

C Heat flux

Assuming that energy radiates uniformly outward from the side and top of the cylindrical flame, the total heat radiation flux released during the combustion of the liquid pool is:

$$Q = \frac{(\pi r^2 + 2\pi r h)\dfrac{dm}{dt}\eta H_c}{72\left(\dfrac{dm}{dt}\right)^{0.61} + 1} \tag{6-52}$$

where Q——Total heat flux, W;

η——Efficiency factor, 0.13—0.35.

The remaining symbols have the same meaning as before.

D Incident thermal radiation intensity of the target

Assuming that all radiant heat radiates from the small sphere at the center point of the liquid pool, the incident radiation intensity measured at a distance(X)from the center of the pool is:

$$I = \frac{Q t_c}{4\pi X^2} \tag{6-53}$$

where I——Intensity of thermal radiation, W/m^2;

Q——Total heat flux, W;

t_c——The coefficient of heat conductivity, in the absence of relatively ideal data, can be taken as 1;

X——Distance from the target point to the center of the pool, m.

of course, there are other factors should be considered.

Air transmittance τ:

$$\tau = 2.02\,(p_w x)^{-0.09} \tag{6-54}$$

where p_w——the partial pressure of water vapor, Pa;

x——the distance between the flame and the receiving surface, m.

$$I_0 = I F_{vg} \tau \tag{6-55}$$

6.4.3.2 Jet fire damage model

A jet is formed when a pressurized combustible material leaks, and a jet fire is formed if it ignites at the leakage port, as shown in Figure 6-12. The method used here for calculating the radiant heat of a jet fire is an extension of the jet diffusion pattern including the airflow effect. Thinking of the entire jet fire as consisting of all the point heat sources along the Central Line of the jet, each of which has the same heat flux.

The heat flux of the point heat source is calculated by the following formula:

$$q = \eta Q_0 H_c \tag{6-56}$$

where q——The heat flux of the point reservoir, W;

η——Efficiency factor, 0.13—0.35;

Q_0——Leakage rate, kg/s;

H_c——Heat of combustion, J/kg.

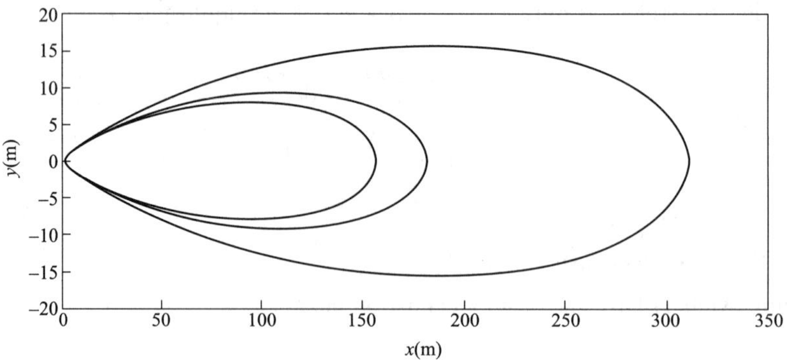

Figure 6-12 Jet fire damage model

The flame length of the jet is equal to the length of the jet axis from the nozzle to the lower limit of combustion of the combustible mixture(LFL).

For the surface flame heat flux, it is concentrated at LFL/1.5.

The division of n points can be arbitrary, and the risk assessment analysis is generally $n=5$.

The heat radiation intensity between a certain heat source I on the jet axis and a point x away from the store is:

$$I_i = \frac{qR}{4\pi x^2} \tag{6-57}$$

where I_i——The intensity of the heat radiation from the point reservoir i to target x, W/m^2;

q——The radiant flux of the point heat source, W;

x——The distance from the point reservoir to the target, m;

R——Emissivity, it depends on the nature of the burning material, 0.2.

The incident thermal radiation intensity at a target is equal to the sum of the thermal radiation intensity of all the point heat sources of the jet fire:

$$I = \sum_{i=1}^{n} I_i \tag{6-58}$$

where n——The number of reservoirs you pick at the point, $n=5$.

6.4.3.3 Fireball damage model

Low temperature flammable liquid gas due to overheating, the pressure of the container increases, is the container explosion, the content is released and ignited violent combustion, producing a powerful fireball, the formation of a strong radiation.

A The fireball radius

$$R = 2.665M^{0.327} \tag{6-59}$$

where R——The fireball radius, m;

M——The mass of rapidly evaporating combustible matter, kg.

B Fireball duration

$$t = 1.089M^{0.327} \tag{6-60}$$

where t——Fireball duration, s.

C The radiant heat flux released when a fireball burns

$$Q = \frac{\eta H_c M}{t} \tag{6-61}$$

where Q——Radiant heat flux during combustion of a fireball, W;

H_c——Heat of combustion, J/kg;

η——The efficiency factor, which depends on the vapor pressure p of the fuel in the container, $\eta = 0.27p_{0.32}$;

t——Fireball duration, s.

D The intensity of incident thermal radiation received by the target

$$I = \frac{QT_c}{4\pi x^2} \tag{6-62}$$

where T_c——Conduction coefficient, $T_c = 1$;

x——The horizontal distance of the target from the center of the fireball, m.

6.4.3.4 Solid fire

The thermal radiation parameters of solid fire are estimated by point source model.

The model assumes that the energy emitted by the flame is part of the combustion, and the radiation intensity is inversely proportional to the square of the center distance of the target fire source:

$$q_r = \frac{fM_c H_c}{4\pi x^2} \tag{6-63}$$

where q_r——The intensity of radiation received by the target, W/m;

f——The radiation coefficient, $f = 0.25$;

M_c——Burn rate, kg/s;

H_c——Heat of combustion, J/kg;

x——Horizontal distance from target to fire center, m.

6.4.3.5 Sudden fire

The leakage of combustible gas, liquid vaporized steam in the air diffusion, meet the source of a sudden combustion without explosion. In this case, all indoor personnel within the range of gas combustion will be burned to death; Some of the people in the building will be burned to death.

The analysis of the consequences of sudden fire is mainly to determine the boundary of the upper and lower limits of the combustion of combustible mixed gas and the range of its lower limit with the diffusion of air mass. Therefore, air mass size and density of combustible mixture can be calculated according to air mass diffusion model.

6.4.3.6 Indoor fire damage model

Estimating by the method recommended by The Architectural Society of Japan.

Evacuation time:

$$t_p = \frac{\rho A_f}{b_p B_p} \tag{6-64}$$

The number of deaths:

$$D_p = b_p B_p (t_p - t_s) \tag{6-65}$$

where t_p——Evacuation time, s;

ρ——Indoor personnel density, person/m^2;

A_f——Room area, m^2;

b_p——Coefficient of evacuation passage, 1.3-1.5;

B_p——Width at narrowest point of evacuation passage, m.

The relationship between the size of the house and the permissible aisle evacuation is shown in Table 6-18.

Table 6-18 The relationship between the size of the house and the permissible aisle evacuation time

Building area(m^2)	Allow aisle evacuation time(s)	Building area(m^2)	Allow aisle evacuation time(s)
<200	30	1000-1500	90
200-500	45	1500-3000	150
500-1000	60		

Property damage:

$$C = A_f P \tag{6-66}$$

where P——The average density of indoor property, Ten thousand yuan/m^2.

6.4.3.7 Fire loss

No matter how the fire burns, it affects the surrounding environment by radiant heat. When the heat radiation intensity is large enough, it can cause the surrounding objects to burn or deform. The intense heat radiation may burn down equipment and even cause casualties. Table 6-19 reveals the loss caused by different incident of thermal radiation.

Table 6-19 Loss caused by different incident fluxes of thermal radiation

The incident flux(kW/m^2)	Damage to equipment	Harm to people
37.5	All operating equipment is damaged	1% dead/10s 100% dead/1min
25	The minimum amount of energy required for wood to burn under a flame-free, prolonged radiation	Serious burns/10s 50% dead/1min
12.5	When there is a flame, the wood burns and the plastic melt at the lowest energy	1 degree burns/10s 1% dead/1min

Continued Table 6-19

The incident flux(kW/m^2)	Damage to equipment	Harm to people
4.0		Feel pain for 20s or more, not necessarily blistering
1.6		Long-term radiation is not comfortable

Comparison of several fire models.

Pool fire:

$$I = \frac{Qt_c}{4\pi x^2} \tag{6-67}$$

Jet fire:

$$I_i = \frac{qR}{4\pi x^2} \tag{6-68}$$

Fireball and deflagration:

$$I = \frac{QT_c}{4\pi x^2} \tag{6-69}$$

Solid fire:

$$q_r = \frac{fM_c H_c}{4\pi x^2} \tag{6-70}$$

6.5 Evaluation method for flammable, explosive and toxic major hazards

6.5.1 Overview

It is an analysis and evaluation method proposed by the Key Project of the Eighth Five-Year Plan "Research on identification and evaluation technology of flammable, explosive and toxic major hazards". It is based on the statistical analysis of a large number of major fires, explosion, poison leakage and poisoning accidents, starting from the material hazard and process hazard, to analyze the causes and conditions of major accidents, evaluate the impact scope, casualties and economic losses of the accidents, and put forward the prevention Control measures. It can accurately evaluate the dangerous degree and level of dangerous substances and processes in the system, and accurately calculate the severity of accident consequences. 107 indicators of process equipment, personnel quality and safety management constitute the evaluation set.

Major hazard installations for dangerous chemicals refer to the hazardous chemicals that produces, processes, uses or stores for a long time or temporarily, and the unit that the quantity of hazardous chemicals is equal to or exceeds the critical quantity.

Unit refers to one(set) production device, facility or place, or several(sets) production devices, facilities or places belonging to the same production and business unit with an edge distance of less than 500m.

Hazardous chemicals refer to chemicals that are flammable, explosive, toxic, harmful and so on, and will cause harm or damage to personnel, facilities and the environment.

6.5.2 Evaluation method

6.5.2.1 Division of evaluation units

Generally, an independent part of the device is called a unit, and each unit is required to have certain functional characteristics. Devices in a common plant can be divided into one unit. All storage tanks in the same dam can be divided into one unit. Scattered pipelines on the ground are not treated as independent units, with the exception of piping bridges.

6.5.2.2 Hierarchy structure of evaluation model

According to the general principles of safety engineering, hazard is defined as the product of accident frequency and accident consequence severity. That is, hazard evaluation depends on the susceptibility of accidents on the one hand, and the severity of consequences once accidents occur on the other hand.

The actual risk not only depends on the inherent risk of the production unit determined by the specific material risk of the production material and the specific process risk of the production process, but also is closely related to various human management factors and the comprehensive effect of disaster prevention and eradication measures.

The evaluation index system of major flammable, explosive and toxic hazard sources can be shown in Figure 6-13.

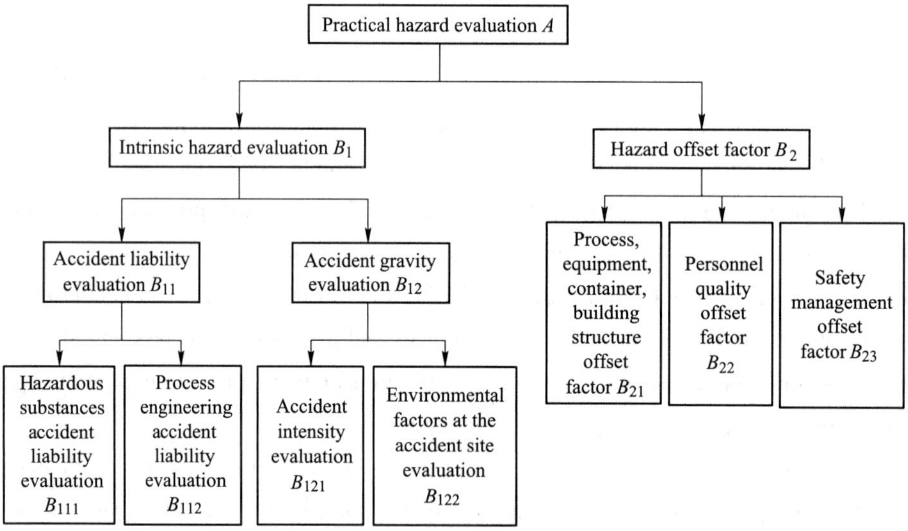

Figure 6-13 The evaluation index system of major hazard sources

6.5.2.3 Mathematical model of evaluation

It is divided into inherent hazard evaluation and practical hazard evaluation.

Inherent hazard evaluation mainly reflects the inherent characteristics of substances, the characteristics of the production process of hazardous substances, and the internal and external environmental conditions of hazardous units.

Practical hazard evaluation is to consider various risk counteraction factors on the basis of inherent danger, which reflects the subjective and dynamic role of people in controlling the occurrence of accidents and the expansion of accident consequences.

Inherent hazard evaluation is divided into accident liability evaluation and accident gravity evaluation.

Accident liability depends on the coupling of accident liability of hazardous substances and process engineering hazards.

The mathematical model for the practical hazard evaluation index of the hazard source:

$$A = \left\{\sum_{i=1}^{n}\sum_{j=1}^{m}(B_{111})_i W_{ij}(B_{112})_j\right\} \times B_{12} \times \prod_{k}^{3}(1 - B_{2k}) \qquad (6\text{-}71)$$

where $(B_{111})_i$—— Evaluation value of the hazard of the i_{th} substance;

$(B_{112})_j$—— Evaluation value of the j_{th} process hazard;

W_{ij}—— Correlation coefficient between the j_{th} process and the i_{th} material hazard;

B_{12}—— Accident severity evaluation value;

B_{2k}—— $k = 1, 2, 3$. B_{21} is the offset factor of process, equipment, container and building structure; B_{22} is the offset factor of personnel quality; B_{23} is the Safety Management offset factor.

A Hazardous substances accident liability evaluation B_{111}

The accident susceptibility of combustion, explosion and toxic hazardous substances is divided into 8 categories.

Each kind of substance is given a weight score (α_i) according to its overall risk sensitivity, which can be found in Table 6-20.

Table 6-20 The weight score according to hazardous substances' overall risk sensitivity

Hazardous substances	α_i
Explosive	1.0
Gas combustibility	1.0
Liquid combustibility	0.9
Solid combustibility	0.5
Spontaneous combustion	0.6
Flammability in water	0.2
Oxidizability	0.3
Toxicity	1.0

The state score (G) of each substance is given according to its physical and chemical parameters related to reaction sensitivity.

Under each major category of substances, there are several sub categories, a total of 19 sub categories. For each major category or subcategory, the scoring standard of the state score is given respectively. The product of the weight score and the state score is the evaluation value of the hazard sensitivity of this kind of substances, that is, the scoring value of the accident susceptibility of hazardous substances (B_{111}). The maximum score of material hazard is 100 points.

$$(B_{111})_i = \alpha_i \times G \tag{6-72}$$

a Explosive

The accident susceptibility of explosives mainly depends on the minimum initiation energy required.

When classifying, it is divided into five levels according to its sensitivity and the combustion and explosion characteristics of its products.

The classification of substances with overall explosion hazard (i. e. level 1.1 and 1.5) or combustion hazard and local explosion hazard (i. e., 1.3) is mainly based on the sensitivity of explosives to thermal, mechanical, electrical and shock waves.

$$G = K_{\mathrm{th}} + K_{\mathrm{e}} + K_{\mathrm{m}} + K_{\mathrm{d}} \tag{6-73}$$

Where, K_{th}, K_{e}, K_{m} and K_{d} are the sensitivity of explosives to thermal, electrical, mechanical and shock waves respectively. The total score of G value is 100 points.

Materials with burst danger (i. e. level 1.2) can be evaluated according to the caliber of ammunition and the sensitivity of agents. The total score of such materials is 80 points.

The sensitivity of the agent to the projectile is graded and evaluated by the critical velocity of the bullet of the initiating explosive. The total score is 30 points.

Materials with severe combustion hazard (level 1.4) can be graded and evaluated according to their combustion speed and spontaneous combustion point, with a total score of 60 points.

b Gas combustibility

Combustible gases are divided into two categories according to their dangerous properties: explosive gases (such as H_2, C_2H_2, LPG, etc.) and combustion supporting gases (such as O_2, compressed air, etc.)

The classification criteria of explosive gases are shown in the Figure 6-14, with a total score of 90 points, including the maximum safety gap, explosion limit, minimum ignition current, minimum ignition energy and ignition temperature.

The classification of explosion limit is carried out according to the explosion risk H:

$$H = \frac{C_{\mathrm{up}} - C_{\mathrm{lower}}}{C_{\mathrm{lower}}} \tag{6-74}$$

There is no evaluation method for combustibility of combustion supporting gas.

Considering the explosion risk of compressed gas containers, the critical pressure and critical temperature are used for classification and evaluation.

When the critical temperature is lower than 21℃, the score is 10 points.

When the critical temperature is between 21℃ and 50℃, the score is 6 points.

c Liquid combustibility

It refers to flammable liquids, liquid mixtures and liquids containing solid substances. Its closed

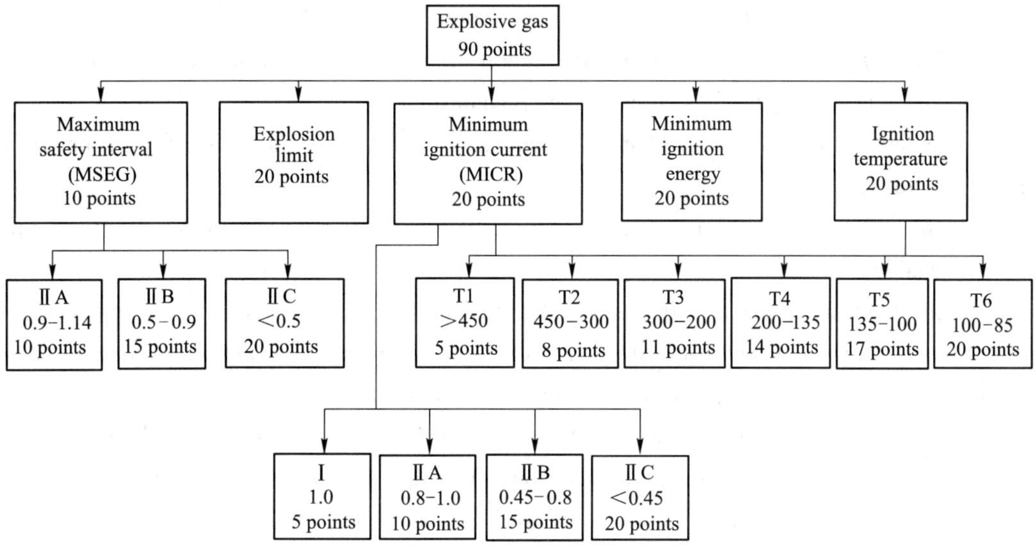

Figure 6-14 The classification criteria of explosive gases

cup flash point ≤61℃.

It is classified according to the flash point and boiling point, and is divided into grade Ⅰ, grade Ⅱ and grade Ⅲ.

Considering its chemical activity, it can be modified on the basis of the original hazard factor B_{111}, and the correction factor is shown in the Table 6-21.

Table 6-21 The correction factor

Chemical activity	4	3	2	1
The correction factor K	0.20	0.12	0.06	0.02

The revised liquid flammability hazard factor is calculated as:

$$B_{111} = B_{111}(1 + K) \tag{6-75}$$

d Solid combustibility

It refers to combustible solid substances other than solid explosives, including flammable solids (level 4.1) and explosive dust (level 4.2).

Level 4.1 flammable solid: classified according to the ignitability and intensity of combustion, with a total score of 20 points.

Level 4.2 explosive dust: it is divided into general industrial dust (requiring external oxygen supply) and self oxygen supply dust (explosive dust, etc.).

The total score of general industrial dust evaluation is 60 points, including specific resistance, ignition temperature of dust cloud, ignition temperature of dust layer, minimum ignition energy of dust cloud and lower limit concentration.

The total score of self oxygenated dust evaluation is 60 points, which is divided into ignition temperature, minimum ignition energy and lower limit concentration.

e Spontaneous combustion

It refers to the substance that does not need the action of external fire source, oxidizes itself with air or is affected by external temperature and humidity, and generates heat and accumulates heat to reach the spontaneous combustion point, causing combustion.

It is divided into primary spontaneous combustion substances and secondary spontaneous combustion substances.

Taking the spontaneous combustion point as the criterion, the total evaluation score is 60 points.

f Flammability in water

This kind of substance reacts with water and releases combustible gas, which will spontaneously ignite without ignition or cocoa combustible gas will ignite, burn and explode, such as alkali metals such as potassium and sodium, hydrides, organic metal compounds, etc.

The generated gas can spontaneously ignite or ignite when it is close to the small flame, or the maximum amount of combustible gas produced by 1kg sample is more than 0. $12m^3/h$.

It is divided into primary water contact combustibles and secondary water contact combustibles.

The total score is 60 points. 60 points for Level 1 and 40 points for Level 2.

g Oxidizability

It is divided into organic peroxides and inorganic oxides (oxidants).

Inorganic oxides can be divided into primary and secondary levels according to oxidation intensity. Substances containing peroxy or high valence chlorine, nitrogen, manganese and other elements, which are easy to obtain electrons and release a large amount of oxygen and heat, are rated as grade 1, with a score of 40. Those with weak oxidizability such as nitrate are rated as grade 2, with a score of 20.

Organic peroxides are easy to decompose and are extremely sensitive to heat, vibration and friction. They are scored according to flammability, mechanical height and automatic accelerated decomposition temperature.

h Toxicity

Susceptibility to poisoning accidents:

$$B_{1118} = B_{1118-1} + B_{1118-2} + B_{1118-3} + B_{1118-4} \tag{6-76}$$

Including material toxicity coefficient, material severity correction coefficient, material odor correction coefficient and material state correction coefficient.

B Process engineering accident liability evaluation B_{112}

The influencing factors of B_{112} process engineering accident liability are determined as 21: Exothermic reaction; Endothermic reaction; Material handling; Material storage; Operation mode; Dust explosion; Low temperature condition; High temperature condition; Negative pressure; High pressure condition; Combustion range and operation nearby; Corrosion; Leakage; Equipment factors; Closed unit; Process layout; Flame; Friction and impact; Pyrosomes; Electrical spark and Static electricity.

Notes:

(1) Exothermic and endothermic coefficients are only applicable to chemical reaction units.

(2) The high temperature coefficient is determined with reference to the melting point, flash point, boiling point and ignition point.

(3) The negative pressure coefficient is only used when the vacuum degree is greater than 500mmHg(66.61kPa).

(4) Correction of high pressure coefficient.

(5) Corrosion rate refers to the sum of internal corrosion rate and external corrosion rate.

(6) The equipment coefficient considers pressure vessels.

(7) Closed unit.

Toxicity Coefficient of process engineering: Corrosion coefficient; Leakage coefficient; Equipment layout factor; Medium influence coefficient; Discharge coefficient; Conveying coefficient; Analysis coefficient.

Notes:

(1) If the corrosion coefficient, leakage coefficient and equipment layout factor have been involved in the fire and explosion hazard assessment, they will not be considered in the toxicity assessment of the process.

(2) Conveying coefficient: Gas delivery, shielded pump, submerged pump.

(3) Analysis coefficient: No ventilation, indoor sampling pipe.

Susceptibility of fire and explosion accidents in process engineering:

$$B_{112} = \frac{100 + \sum_{i=1}^{m} B_{112-i}}{100} \tag{6-77}$$

Susceptibility of poisoning accidents in process engineering:

$$B_{112} = \frac{\left(100 + \sum_{i=1}^{m} B_{112-i}\right)\left(100 + \sum_{j=1}^{n} B_{112-j}\right)}{10\ 000} \tag{6-78}$$

Process-material hazard correlation coefficient W_{ij}:

The combination of process and substance can reflect different risks, and the correlation coefficient is divided into 5 levels, as shown in Table 6-22.

Table 6-22 Process-material hazard correlation coefficient

Level	Correlation	Process-material hazard correlation coefficient W_{ij}
A	Close	0.9
B	Great	0.7
C	General	0.5
D	Small	0.2
E	No	0

C Evaluation method of accident severity

The severity of the accident is expressed by the economic loss of the accident consequences. It refers to casualties and property losses of houses, equipment and materials in the accident, regardless of shutdown losses.

Correspondence between dangerous objects and injury models, see Table 6-23: Maximum risk principle; probability summation principle.

Table 6-23 Correspondence between dangerous objects and injury models

Classification number of dangerous objects	Injury models
1. 1-1. 5, 7. 1, 7. 2	Explosion damage model of condensed energetic materials
2. 1	(1) Gaseous storage is vapor cloud explosion injury model; (2) liquid storage is calculated as $S=AS_1+(1-A)S_2$
3. 1-3. 3	Pool fire model
4. 1, 4. 2, 6. 1, 6. 2	Solid and dust fire injury model

Treatment method of coexistence of multiple dangerous substances in a dangerous unit:

(1) Non explosive substance. Calculate the total loss of each substance in case of accident, and then take the maximum value as the total loss of the unit:

$$S = \max_{1 \leqslant i \leqslant N} (S_i) \tag{6-79}$$

(2) Explosive hazards. Calculate the total explosion energy, and then calculate the total loss according to the total explosion energy:

$$E = \sum_{i=1}^{K} Q_{B,i} W_i \tag{6-80}$$

In case of ground explosion, multiply by the ground explosion coefficient of 1. 8.

Do not consider the interaction between units.

a Injury models

(1) Explosion damage model of condensed energetic materials.

Shock wave damage failure criterion:

Overpressure criterion: only related to overpressure, not related to action time. Applicable when $\omega T^+ > 40$.

Impulse criterion: consider overpressure and action time at the same time. Applicable when $\omega T^+ < 0.4$.

Overpressure impulse criterion: the damage effect is determined by overpressure and impulse

$$(\Delta p - p_{cr})(I - I_{cr}) = C \tag{6-81}$$

When estimating the radius of the dead zone, the overpressure impulse criterion is used; Use overpressure criteria when estimating the radius of serious and minor injuries.

(2) Damage model of vapor cloud explosion.

(3) Fire injury model.

(4) Pool fire damage model.

(5) Boiling liquid diffusion is a vapor cloud explosion (BLEVE) damage model.

(6) Solid fire injury model.

(7) Indoor fire injury model.

b Basic assumptions

The injury or damage of the accident is isotropic.

The death probability of personnel in the death zone is 50%, the eardrum of personnel in the serious injury zone is 50% broken (explosion model) or 50% second degree burn (fire model), and the eardrum of personnel in the minor injury zone is 1% broken (explosion model) or 50% first degree burn (non explosion model).

The radius of property loss is the radius of damage Level 2.

Strictly distinguish between inside and outside the injury area.

Indirect losses are not considered.

The effect of accident prevention measures is not considered.

Death toll:

$$N_1 = 3.14\rho_1(R_{0.5}^2 - R_0^2) \tag{6-82}$$

Number of serious injuries:

$$N_2 = 3.14\rho_2(Rd_{0.5}^2 - R_{0.5}^2) \tag{6-83}$$

Number of minor injuries:

$$N_3 = 3.14\rho_3(Rd_{0.01}^2 - Rd_{0.5}^2) \tag{6-84}$$

Property loss:

$$C = 3.14\rho_4 R_4^2 \tag{6-85}$$

Accident risk:

$$S = C + 20\left(N_1 + \frac{N_2}{2} + \frac{105}{6000N_3}\right) \tag{6-86}$$

D Hazard offset factor

(1) Process equipment and buildings: fire and explosion, toxicity and poisoning prevention, with 28 indicators;

(2) Quality of post operators: 4 indicators;

(3) Safety management status: 10 indicators.

(4) Offset factors of process equipment and buildings: Equipment maintenance; Explosion suppression device; Inert gas protection; Emergency cooling; Emergency supply; Electrical explosion proof; Antistatic; Lightning protection; Fire arrestor; Accident discharge and treatment; Device monitoring; Equipment layout; Process parameter control; Leak detection device and response; Fault alarm and control device; Plant ventilation; Building pressure relief; Plant structure; Industrial sewer; Fire resistant support; Storage tank; Workshop; Isolation operation; Poison detection; Emergency damage system; Personal protective equipment; Wind vane and First aid for poisoning.

(5) Quality of dangerous post operators: Personnel qualification; Proficiency; Stability and Load factor.

(6) Safety management: Safety production responsibility system; Safety production education;

Safety technical measures plan; Security check; Safety production rules and regulations; Safety production management organization and personnel; Accident statistical analysis; Hazard source evaluation and rectification; Emergency plan and measures and Fire safety management.

6. 5. 2. 4 Hazard classification and hazard control degree classification

A The unit risk is classified based on its inherent risk

$$A^{*} = \lg(B_1^{*}) \tag{6-87}$$

The classitication of hazard is shown in Table 6-24.

Table 6-24 Hazard classification

Major hazard level	Ⅰ	Ⅱ	Ⅲ	Ⅳ
A^{*}/¥ 100 000	≥3. 5	2. 5-3. 5	1. 5-2. 5	<1. 5

Level Ⅰ : National supervision;

Level Ⅱ : Provincial and municipal supervision;

Level Ⅲ : County and city supervision;

Level Ⅳ : Enterprise supervision.

B Hazard source control level

The smaller the value B_2 of the comprehensive offset factor, the higher the degree of control, which can be found in Table 6-25.

Table 6-25 Hazard source control level

Unit hazard control level	A	B	C	D
B_2	<0. 001	0. 001-0. 01	0. 01-0. 1	>0. 1

Level Ⅰ : Above grade A;

Level Ⅱ : Above grade B;

Level Ⅲ : Above grade C;

Level Ⅳ : Above grade D.

6. 6 Modern safety management methods

6. 6. 1 PDCA cycle

PDCA cycle was first proposed by Walter A. Shewhart, an American quality management expert. It was adopted, publicized and popularized by Dai Ming later, so it is also called Dai Ming Ring. The ideological basis and method basis of total quality management is PDCA cycle. PDCA cycle divides the quality management into four stages: Plan, Do, Check and Act, which is shown in Figure 6-15. During the quality management activities, each work is required to plan, implement and check the implementation effect. The successful work will be included into the standard, while the unsuccessful will be leaved to be solved in the next cycle. PDCA cycle is the basic method of quality management and the general rule of enterprise management.

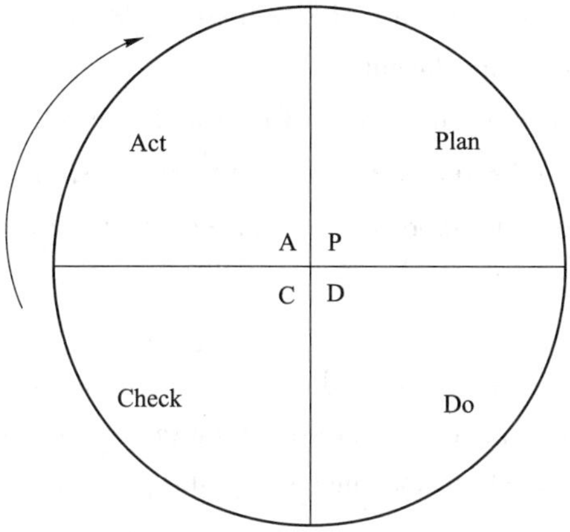

Figure 6-15 The PDCA cycle

6.6.1.1 The meaning of PDCA

PDCA is the first letter of the English words Plan, Do, Check and Act, see Figure 6-16. PDCA cycle is a scientific procedure that carries out quality management in this order and goes on continuously.

(1) P: Plan. The determination of policies and objectives, and the formulation of activity planning.

(2) D: Do. Design specific methods, schemes and plan layout based on known information. Then, according to the design and layout, carry out specific operations to achieve the contents of the plan.

(3) C: Check. Summarize the results of the implementation plan. Distinguish what is right and what is wrong, clarify the effect, and find out the problems.

(4) A: Act. Process the results of summary inspection, confirm the successful experience, and standardize. We should also sum up and pay attention to the lessons of failure. The problems not solved shall be submitted to the next PDCA cycle for solution.

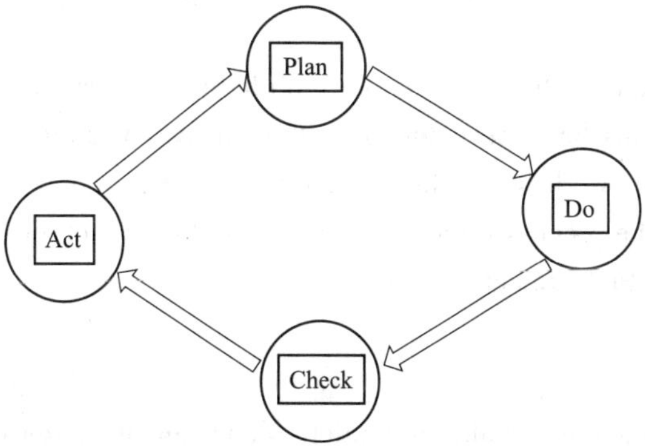

Figure 6-16 Deming Model

6.6.1.2 Modern viewpoint

(1)P:Planning. The planning function includes three parts:goal,plan and budget.

(2)D:Design. Design scheme and layout.

(3)C:4C management:Check,Communicate,Clean and Control.

(4)A:2A. Act(execute,handle the results of summary inspection) and Aim(act according to the objective requirements,such as improvement and enhancement).

6.6.1.3 How to apply

(1) Planning phase. It is necessary to find out the requirements of users for product quality through market research,user visits,etc. Then determine quality policies,quality objectives,quality plans,etc. Including the investigation and analysis of the current situation,determining the main causes and making plans.

(2)Design and implementation phase. Implement the contents specified in the previous stage. Conduct product design, trial production, test and personnel training before plan implementation according to quality standards.

(3)Inspection phase. It is mainly to check the implementation during or after the implementation of the plan to see whether it meets the expected results of the plan.

(4) Processing phase. It is mainly based on the inspection results to take corresponding measures. Consolidate the achievements,incorporate the successful experience into the standard as much as possible,and carry out standardization. The remaining problems will be transferred to the next PDCA cycle to solve.

The above four processes do not end once, but go on again and again. After one cycle is completed,some problems are solved. Then the unsolved problems enter the next cycle. In this way, the problem rises step by step. PDCA cycle is a scientific procedure for total quality management. The whole process of total quality management activities is the process of formulating and organizing the implementation of the quality plan. This process is run in a PDCA cycle without stopping. The four stages of PDCA cycle, the PDCA cycle management mode both embody a specific management method and a set of scientific working procedures of scientific epistemology. The application of PDCA management mode has great benefits for us to improve the efficiency of daily work. It can be used not only in quality management, but also in other management work.

But PDCA does not contain human creativity. With the application of PDCA in more project management,many problems have been found in the process of application. It just taught people how to improve the existing work,which led to inertial thinking. People who are used to PDCA can easily work according to the process without creativity due to the lack of pressure. Therefore,PDCA has some limitations in actual projects.

6.6.2 OHSMS

The abbreviation of Occupational Health and Safety Management Systems is "OHSMS",which is a modern safety production management mode that rose in the world in the late 1980s. Together with

ISO9000, ISO14000 and other standard systems, it is called "the management method in the post industrial era". This system is formulated to prevent occupational disease hazards, protect workers' health, enhance employees' awareness of safe production and ensure production safety. The main reason for its emergence is the requirements of the enterprise's own development. With the expansion of enterprise scale and the improvement of production intensification, higher requirements are put forward for enterprise quality management and business model. Enterprises must adopt modern management mode to make all production and operation activities, including safety production management, scientific, standardized and legalized.

6.6.2.1 The characteristics of OHSMS

(1) Control occupational health and safety performance by establishing a management system;

(2) Adopt PDCA cycle management;

(3) Emphasize prevention first, continuous improvement and dynamic management;

(4) Compliance with the requirements of laws and regulations runs through the system;

(5) All staff are required to participate;

(6) Applicable to different walks of life and serves as the basis for certification.

6.6.2.2 Management role of basic elements of OHSMS

(1) The planning of hazard identification, risk assessment and risk control is the basis of OHSMS;

(2) It has the function of realizing the commitment to comply with regulatory requirements;

(3) The monitoring system of it guarantees the system operation;

(4) Clarifying the organizational structure and responsibilities is a necessary prerequisite for implementing the OHSMS;

(5) Other OHSMS elements have unique management functions.

6.6.2.3 Specific work methods

Specific work methods of occupational health and safety management system can be summarized in six words: say, do, remember, check, change and verify.

(1) Say: explain the policy of the organization;

(2) Do: implement according to the requirements of the policy;

(3) Record: record the implementation details;

(4) Check: check the implementation and record the implementation;

(5) Correction: timely rectify the problems occurred in the implementation process;

(6) Inspection: timely track and verify the rectification.

6.6.3 HSE management system

The HSE management system refers to the whole of the organization, responsibilities, practices, procedures, processes, and resources for the implementation of safety, environment, and health management, which is shown in Figure 6-17. H-health refers to a person who is physically free from diseases and remains in a psychologically intact state; S-safety refers to efforts to improve working conditions and overcome unsafe factors in the process of labor production, so that labor production can ensure the health of workers and enterprise property. E-environment refers to the sum of various natural forces or functions that are closely related to human beings and affect human life and production activities, including the combination of various natural factors and human combination of ecological relationships formed with natural factors. Elements are organically integrated through an advanced and scientific operation mode, interrelated, and interact to form a set of structured dynamic management system. Its function is to carry out risk analysis in advance, to determine the possible hazards and consequences of its own activities, and to take effective preventive measures and control measures to prevent it from happening, so as to reduce possible personal injury, property loss and effective management of environmental pollution. model. It highlights the prevention and continuous improvement in advance, and has a high degree of self-discipline, self-improvement, and self-motivation mechanism. It is a set of health, safety and environmental management system commonly used in the international petroleum industry for oil and gas exploration and development and construction industry. It is also a management mode widely recognized by international petroleum and petrochemical industry companies, which is characterized by systematization, scientization, standardization and institutionalization. Therefore, the HSE management system is a modern management model and one of the modern enterprise systems. And it is inextricably linked in the actual work process. It is inevitable for modern petrochemical enterprises to form an integrated management system for Health, Safety and Environment.

Figure 6-17 The PDCA cycle

To successfully establish and operate an HSE management system, one must first have a deep understanding of the basic elements of HSE management system standards. Based on the SY/T 6276—1997 standard and with reference to the " Health, Safety and Environmental Management System Management Manual" of China National Petroleum Corporation, the following describes the

7 elements of the management system standard in turn, and the 7 key elements of the HSE management system. Shown in Figure 6-18.

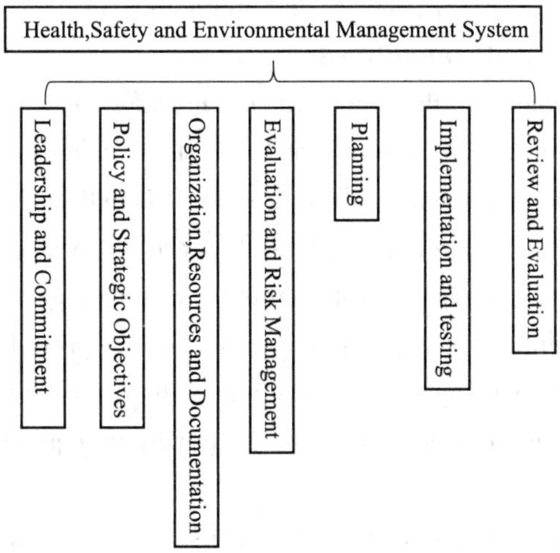

Figure 6-18 Key elements of HSE management system

Before explaining each element, it should be specially pointed out: The SY/T 6276—1997 standard expresses the necessary elements for establishing, implementing, and maintaining a health, safety, and environmental management system, and is a comprehensive and comprehensive management framework. The purpose of this system standard is to help the company itself and related parties (such as contractors and suppliers) to achieve health, safety, and environmental management goals, rather than to replace the company's existing, effective, and feasible management methods and systems. This standard does not prescribe specific performance criteria but recommends that companies consider the significant hazards and impacts that a company's activities may have when setting policies and objectives.

The system standard is divided into 7 first-level elements and multiple second-level elements. The purpose of this division is to facilitate expression and understanding, but in fact these elements are closely related, that is, one element may sometimes involve several other elements. Elements, the change of any one element must consider the impact on other elements at the same time, and each element must not be understood in isolation.

6. 6. 3. 1 Leadership and commitment

Leadership and commitment refer to the leadership and commitment of management at all levels from top to bottom and is the core of the HSE management system. Senior management should provide strong leadership and a clear commitment to responsibility and management of health, safety and the environment, and ensure that leadership and commitment are translated into the resources necessary to establish, operate and maintain the HSE management system and achieve established policies and Strategic goals, through the implementation of leadership commitments, strive to create a corporate culture that makes commitments permanent in the hearts of all employees. China National Petroleum Corporation requires: The top management of the enterprise should make a clear commitment to HSE management and strive to create and maintain a good corporate culture to support the HSE policy of the group company.

A The role of leadership and commitment

The leadership and commitment of the senior managementare the core of the HSE management system and the driving force for the operation of the system, which is of great significance to the establishment, operation, and maintenance of the system. It is self-evident, because no matter what type of management system is adopted, it will be difficult to do without the leadership and support of managers, and leaders at all levels are responsible for leading and mobilizing all staff to achieve the goals and targets of health, safety, and the environment. Responsibility, the role of leadership is to continuously improve the company's HSE performance by demonstrating correct HSE behavior, by identifying HSE responsibilities and obligations, by providing required resources, and by-passing assessments and audits.

The leadership role of senior leaders in the establishment and implementation of HSEMS is reflected in: setting an example of HSE and supporting correct behavior through all-round physical action; clear two-way communication with employees, contractors and other relevant parties on HSE-related issues; The requirements are comprehensively reflected in the business development plan to ensure the establishment of written management requirements; the HSE management system is guaranteed to operate in accordance with the established policies and objectives in terms of ideology, organization and system, taking into account production, business and other aspects; establish clear HSE objectives, standards, responsibilities, performance assessment methods, and equipped with corresponding human and material resources; in the company, according to the annual goals, the leaders of each unit should be assessed, and opinions from all aspects should be solicited during the assessment; the indicators established by the group company should be implemented. To the company's business activities, such as external certification, sustainable development, conservation of biodiversity, etc. ; to promote internal and external exchanges of HSE experience. Senior leaders' commitment to health, safety and environmental responsibility and management comes from the company's initial risk analysis and the company's HSE policy established by the leadership. Therefore, the commitment indicates the direction of HSE work and shows the leadership's determination to do HSE work well, has a huge appeal to the majority of employees. Commitment to HSE should start from the top management of the organization. The top management of the organization should make a clear commitment to the HSE policy of the organization in the form of documents. The commitment of the senior management of the company is a public commitment to all employees and society. The commitment of the top leaders shows that: the top managers have the primary responsibility and obligation to do a good job in health, safety and environment; HSE management is one of the priorities in the company's entire management system, and HSE management should be treated like other management of the company. Guarantee full support for the establishment, implementation and maintenance of the system, such as giving HSE personnel the power they should have, allocating necessary resources, and ensuring the investment and availability of funds and time; strive to create and maintain a corporate culture of HSE management, encourage all High-level employees and contractors jointly participate in the HSE management of the enterprise; establish external communication channels with the government and the public, and regularly announce HSE performance; leaders should set a good example for employees.

The leadership and commitment of the management at all levels, in addition to the above content, is also reflected in: effective communication of HSE content in corporate decision-making; HSE should be given top priority in management meetings, and leaders should attend and preside over HSE meetings and related inspections, Actively participate in local and field HSE activities and reviews; summarize HSE activities upon completion; encourage employees to make suggestions for improving HSE performance; include HSE responsibilities in everyone's assignments; participate in internal and external HSE Innovation.

B Requirements for commitment

It should be proposed and documented by the top leaders before the system is established; the opinions of employees and the society on the commitment should be solicited before the formal presentation; the commitment should be clear and concise, so that employees and the public can easily understand and grasp it; the commitment should be open and transparent, and Use various forms to publicize, such as posting, going on the Internet, etc. ; promise to be deeply rooted in the hearts of the people and become an integral part of the corporate culture; when conditions change, senior leaders should propose revisions and make revisions when the management system management manual is revised.

C Basic content of commitment

Commitment to achieving the safety, health and environmental management system policy, strategic objectives, and plans; commitment to HSE priority and effective implementation of the HSE management system; commitment to meeting HSE requirements and regulations in all activities; expectations of employee HSE performance; expectations of contractor HSE performance.

D Top-down commitment

Personnel at all levels should have a clear understanding of the organization's public commitments and their HSE responsibilities and make corresponding commitments accordingly. The commitments made by managers at all levels and all employees of the organization should be consistent with the organization's policies and goals and consistent with their responsibilities.

E Establish a corporate culture that supports HSEMS

Corporate culture mainly refers to the spiritual wealth formed during the creation and development of an enterprise. HSE corporate culture should be an integral part of the entire corporate culture. Only when an enterprise cultivates a good HSE corporate culture, can the HSE written on paper become the actions of all employees, and finally realize the requirement of " HSE into my heart". HSE culture includes beliefs, values, driving forces, personal commitment, participation and responsibility, etc. , mainly reflected in: belief in continuous improvement of HSE performance; encouragement and promotion of employees to improve HSE performance; each employee's responsibilities and obligations reflect the company's HSE performance; employees at all levels participate in the establishment and operation of the HSE management system; implement the HSE management commitment from top to bottom; ensure the effective implementation of the HSE management system. The establishment process of corporate culture is a process in which the HSE management system changes from paper(document) to action, and then to belief. Leadership has a major responsibility for nurturing this culture, and the company's employees and contractors are involved in the creation and maintenance of the HSE corporate culture.

[Example 1] CNPC's commitment to HSE

China National Petroleum Corporation has always believed that the most important resource in the world is human beings and the natural environment on which human beings depend. Protecting the environment, protecting the health of employees and the safety of life and property is one of the core tasks of the company. In order to obtain and maintain good health, safety and environmental performance, China National Petroleum Corporation solemnly promises to employees, employees, customers and the society: abide by the laws and regulations of the country and local, respect local customs; protect the environment, make reasonable use of resources, committed to sustainable development; adhere to prevention first, pursue the goal of no accident, no injury and no loss; optimize the allocation of human, material and financial resources, and continuously improve HSE management; the top managers at all levels are the first and responsible persons of HSE, HSE performance and performance is an important basis for rewards and punishments, hiring personnel and hiring contractors; implement HSE training, establish and maintain HSE corporate culture; disclose our HSE corporate culture to the society; anywhere in the world, in the business In any field, we have the same attitude towards HSE. All CNPC employees, employees, customers, and contractors are responsible for upholding the company's commitment to health, safety, and the environment.

[Example 2] Shell's commitment to HSE

Within the group companies, we are unanimously committed to: not harming people is our pursuit; all operations must protect the environment; to provide products and services that use raw materials and energy efficiently; under the premise of being consistent with the above principles, develop energy and products, provide services; regularly publish our performance; take a leadership role in promoting the best HSE practices in our industry; manage HSE like any other critical business; foster a corporate culture where commitment is shared by all Shell employees. The purpose of this is for us to achieve a good HSE performance that we are proud of, to win the trust of our customers, shareholders, and the public, to be a good neighbor and to contribute to sustainable development.

[Example 3] Statoil's commitment to HSE

Excellent performance in health, safety and environment has intrinsic value, good and efficient performance is the premise of competition; our goal is no accident, no injury and no loss; we will contribute to sustainable development; We have the same attitude everywhere and in every area of business; expect our suppliers and partners to meet the requirements of our HSE standards.

6.6.3.2 Policy and Strategic Objectives

The guiding ideology and code of conduct for HSE managementformulated by senior leaders for the company and are the intentions and principles of action for health, safety, and environmental management. The goal of improving HSE performance is the basis and guide for the establishment and operation of the system. Health, safety, and environmental management are inseparable, and the HSE policies formulated should not be independent of each other but should be comprehensive. The HSE management policy of China National Petroleum Corporation is people-oriented, prevention-oriented, leadership commitment, full participation, system management, and continuous improvement. Level up. The significance of formulating policies and strategic objectives lies in the following: the policies and strategic objectives determine a general guiding direction for the enterprise and are the guiding ideology and code of conduct for carrying out management work, which should be reflected in the management objectives and plans at all levels of the enterprise. The policy and strategic objectives stipulate the goals pursued by the enterprise in HSE management and the direction and approach to be followed to achieve this goal, and are the basis for the enterprise to establish specific HSE goals and indicators.

The policies and strategic objectives are formulated and issued by the top management of group companies or subsidiaries. The policy and strategic objectives should bedocumented, and the content should be simple, clear, and motivating. The strategic objectives should provide a high-level overview of senior management 's requirements and expectations for HSE.

A Content

The content of policies and strategic objectives should at least include compliance with relevant laws and regulations, and adoption of current standards in areas not specified by laws and regulations; the idea of continuous improvement; emphasis on accident prevention; expectations for company employees and contractors. Require. Policy and strategic objectives may also include content for all aspects of HSE: Health: Create a healthy working environment and actively promote the improvement of employee health and well-being; Safety: Prevent all safety incidents that may occur in company activities; Environment: Progressively reduce emissions, wastewater, and solid waste to ultimately eliminate their adverse impact on the environment. For the subsidiary companies of the group company, they should also include some targeted content when establishing their own policy objectives, such as specific safety, health and environmental protection requirements for fire prevention, explosion protection, and anti-toxicity.

B Requirements to be met

HSE policies and objectives must be consistent with current laws and policies, consistent with other policies and objectives of the enterprise, and reflect the common will of all employees; a decomposable target network can be formed, which can be achieved through the joint efforts of all employees; the objective cannot be If the target is set too high, it cannot be set too low. If the target is too high, it will affect the enthusiasm of employees and damage the corporate image. If the target is too low, it will lose the motivation and motivation of employees. Environmental performance level. Specifically, it should reflect: be coordinated and consistent with other company policies and objectives; have the same importance as other company policies and objectives; be consistent with the parent company's policies and strategic objectives; be implemented and implemented by organizations at all levels; public It is easy to obtain; it meets or exceeds the requirements of relevant laws and regulations; if there are no relevant provisions in local laws and regulations, the company's appropriate standards should be selected or formulated; the impact of activities, products and services on health, safety and the environment should be minimized as effectively as possible. risks and hazards; the operating company's HSE target, the content should be as specific as possible, preferably quantified. For example, determine the starting point and indicators (the starting point for meeting all environmental protection regulations as the guideline, and the statutory standard as the indicator), as well as resource requirements, start and completion times, etc. The enterprise should formulate a series of effective procedures and provide suitable facilities and equipment to ensure the realization of the policy and objectives; the enterprise should have a complete system of information collection, follow-up investigation, and dynamic review to ensure the compliance and effectiveness of the policy objectives.

C Written format

The company's HSE policy should be written in the same format, i. e. : it should be written and printed in an easy-to-read format; it should be in a language familiar to employees; it should be consistent across companies in appearance; the company name and its business scope should be clearly stated; Generally, it should be limited to a piece of A4 or A3 paper, and it can also be made into a small card; it should be signed by the general manager of the company.

D Lifecycle

Because all systems of control become partially or fully obsolete over time and conditions, the HSE policy should be periodically reviewed, focusing on the policy's purpose, scope, and its continuity and adequacy, to check for the need for the control system is modified or supplemented. Dedicated personnel should oversee this work to ensure timely action.

[Example 4] Shell's HSE policy and strategic goals

Each company belonging to Shell must have a set of HSE management system to ensure that it abides by the law in business activities, continuously improves and achieves better performance; continuously proposes improvement goals, measures, evaluates and reports on its own HSE aspects Require contractors to manage HSE affairs and work in accordance with this policy; require joint ventures operating under Shell's control to apply this policy and use their influence to promote this policy in other enterprises; when evaluating employees, including health, safety, and environmental performance, and rewarding caregivers accordingly.

[Example 5] BP-Amoco's HSE policy and strategic goals

Every company employee, no matter where they are, has a responsibility for health, safety, and the environment. Good health, safety and environmental performance is the key to our business success. The company will pursue and achieve good health, safety, and environment (HSE) performance globally in a differentiated approach. The company's commitment to health, safety and environmental performance is one of the Group's five corporate policies(ethical conduct,employees, public relations, health, safety and environmental performance, financial management). We are committed to taking action to demonstrate our value for the natural environment and strive to achieve our goals:no accidents,no harm to human health,and no damage to the environment.

[Example 6] CNPC's HSE Policy and Strategic Objectives Guide

The company shall have clear and documented health,safety and environmental policies and strategic objectives and ensure that they are:consistent with the policies and strategic objectives of any parent company; consistent with the company's activities, products, services and their Environmental impacts are closely related and relevant; are consistent with other company policies; have equal importance with other company policies and objectives; are implemented and implemented at all levels of the organization; easily accessible to the public; meet or exceed the requirements of relevant rules and regulations;local When laws and regulations do not have relevant provisions,the appropriate standards within the company can be used;the risks and hazards to health,safety and the environment caused by activities, products and services can be reduced as effectively as possible;health,safety and environmental goals are established through continuous efforts. Improve health,safety,and environmental performance levels.

The company shall establish and periodically review health,safety,and environmental strategic objectives. This objective should be consistent with company policy and reflect operational and commercial requirements for the activity and related health,safety and environmental hazards and impacts, and reflect the views of employees, contractors, customers, and companies engaged in similar activities.

6.6.3.3 Organization,Resources,and Documents

Organizational structures, resources and documents are the organizational guarantee and material basis for the operation of the system and are necessary conditions to ensure the good performance of HSE. China National Petroleum Corporation HSE management system requirements: strive to achieve optimal configuration in personnel organization, resource management and document management,and implement HSE responsibility management to achieve good HSE performance.

This element contains 7 secondary elements,see Table 6-26.

Table 6-26 Secondary elements of "Organization,Resources and Documents"

Secondary elements	Gist
Organization and responsibilities	Organizational system and responsibilities of personnel at all levels
Management representative	Duties and authorities of management representatives
Resource	Optimal allocation of resources
Ability	Assessment and necessary training of the competencies that must be possessed by employees engaged in the main activities and tasks of HSE
Contractor	Requirements for HSE management of contractors

Continued Table 6-26

Secondary elements	Gist
Information exchange	The common understanding that the company, contractors and collaborators should have on HSE affairs, information exchange
Documents and their controls	Control file content and file management

A Organization and responsibilities

Organizational structure refers to the composition and responsibilities of the departments and personnel responsible for HSE management in the enterprise management system and is the specific management organization of the enterprise HSE management system. Resources mainly refer to the available internal resources such as human, financial, material, technology, equipment, etc. , which are important material guarantee documents for the establishment and operation of the HSE management system. A document that can be written or electronic. The HSE management system involves personnel from different departments and at different levels, and they are given different powers and responsibilities, and the resources required to be allocated are also different. Obligations and interrelationships are clearly defined, documented, and published.

a Organization

According to the current state of the enterprise system in China, it is not feasible for the HSE system organization to copy foreign ones, and it can only be determined according to our specific situation. The role of the HSE organization is to connect the institutions and personnel responsible for HSE affairs to form an organic and combat-effective whole, forming a hierarchical network system. China National Petroleum Corporation operates in a linear management mode, which is consistent with the management system of the group company. The HSE management committee of the enterprise will accept the guidance of the HSE steering committee of the group company. The director of the HSE steering committee is the general manager, and the top management representative is appointed to oversee HSE work. Each level has its own management scope and corresponding HSE management system and organization, and the HSE department at the next level plays a leading and supporting role (mainly supporting and guiding) to the subordinate department.

China National Petroleum Corporation implements alinear organizational system headed by the director of the HSE Steering Committee of the group company and its authorized management representatives and heads of various departments. The director of the HSE steering committee is the general manager, and the executive department is the quality, safety, and environmental protection department. Subsidiary companies and HSE organizations of enterprises Each enterprise has set up an HSE management committee, the director is the top manager, and a special HSE management organization is set up. HSE Organization Project units establish corresponding HSE management organizations, and construction teams establish HSE teams. Only a reasonable organization can ensure that all specific responsibilities are implemented to individuals in relevant departments, and that various issues and responsibilities can be traced back. However, the establishment of the organization is relatively easy, but the transformation of functions, the improvement of the quality of personnel and the transformation of the thinking mode requires long-term efforts. Once the HSE organizational structure is established, the upper-level leaders shall assign responsibilities and authorities, and each organization shall clearly define the individual's post responsibilities and provide necessary resources. Only when the organization is reasonable, everyone performs their own duties and responsibilities, and has resources to ensure that the organization can ensure that the company's HSE activities are carried out in accordance with the prescribed requirements and make correct responses when abnormal situations occur.

b Responsibilities

Responsibilities refer to the specific responsibilities of managers at all levels to establish, implement and maintain the HSE management system within their jurisdiction and the responsibilities of individual employees in ensuring HSE performance. The responsibilities of key agencies and managers responsible for HSE management are briefly described below.

The responsibility of the director of the HSE Steering Committee is to implement the national HSE laws and regulations; be responsible for the overall establishment and implementation of the HSE policy; responsible for making commitments: HSE Steering Committee meetings, decision-making on major issues: providing necessary resources; regularly reviewing the HSE management system Conduct management reviews and provide suggestions for continuous improvement.

The responsibilities of the representative of the HSE top management are to undertake all the responsibilities entrusted by the committee; formulate HSE strategic goals and performance indicators; hold regular meetings of the steering committee to review the HSE work report at the stage of validation; organize and review the annual HSE work plan of the group company, and supervise the implementation; Approval of standards, regulations, practices, and guidelines.

The responsibilities of the HSE executive agency are to organize the establishment and maintenance of HSE management system documents, and be responsible for system operation management; organize the formulation and implementation of the group company's annual HSE plans and plans; formulate HSE management system audit plans, organize HSE management system audit activities; handle HSE management Relevant problems in the operation of the system; responsible for coordinating and solving HSE technical problems, and providing technical services for HSE management of the group company.

Each professional health, safety and environmental management department are to be responsible for the professional and technical management of HSE, to provide technical support for the HSE decision-making of the group company; to organize the supervision, inspection and monitoring of the implementation of the HSE work plan of each directly affiliated unit; Investigation and handling of accidents; organize HSE scientific research and academic exchanges; organize HSE management and technical personnel training, and manage HSE scientific research and consulting institutions affiliated to the group company in a centralized manner.

Because each functional management department has the obligation to support the establishment and operation of the HSE management system and undertake HSE management responsibilities within the corresponding jurisdiction.

The responsibilities of the HSE management committee are to implement the HSE policy of the group company, support and participate in the HSE management review organized by the group company; organize the preparation and revision of the HSE management system documents of the unit, and be responsible for the system operation and management; participate in the comprehensive decision-making of the enterprise, organize Formulate the annual HSE goals, indicators and plans, and authorize the handling of HSE issues of the unit; preside over the internal HSE management system audit work of the unit, and approve the HSE management system audit plan and report; report to the HSE Committee of the group company every year on the completion of various indicators and work implementation If the situation arises, put forward the next work arrangement and submit it to the group company for review and record; organize and carry out HSE publicity, education, training, technical information exchange activities, and be responsible for the contact with external parties on matters related to the HSE management system; accident, establish and improve HSE records, statistical reports and technical files.

The responsibility of the HSE management representative is to be responsible for the overall performance of the HSE; to ensure that all production sites incorporate a detailed HSE management system; to assign responsible persons to each component of the system; to determine and control risks; to improve the ability of continuous improvement of HSE management.

HSE on-site supervision are to be responsible for interpreting relevant regulations, measures and HSE guidelines, formulating and writing technical standards, practices and guidelines for the unit; monitoring and evaluating HSE performance, updating and maintaining records, and writing HSE performance stage reports; ensuring that emergencies occur Have a good communication system from time to time to make effective emergency response; complete independent audits to assure L-level authorities that all risks can be effectively controlled; learn lessons from internal or external accidents in the group company, and master the latest technology through various channels, to promote HSE performance; solicit expert opinions for various activities of the unit, organize the development of new HSE technologies and processes; provide technical and engineering rationalization suggestions for superior leaders to formulate HSE strategies and plans; organize and carry out HSE risk assessments; Monitor the HSE performance and production activities of the business.

Employees have the obligation and right to maintain the HSE corporate culture. They need to receive HSE training, complete work tasks in accordance with HSE documents, accept inspection and supervision by the HSE management department, report accidents, including near-miss accidents, and participate in HSE management.

B Management representative

The top management is often busy with official duties, and it is impossible to be personally responsible for HSE management. Therefore, a representative of the top management must be authorized to perform necessary responsibilities. The top management representative can manage on behalf of the top management within the scope of authorization but should report to the top management and ask for instructions on a regular basis. The management representative is generally the first deputy of the top management. Top management should allocate the resources needed to exercise the authority of the HSE management representative. For the responsibilities of the top management representative of the group company and the management representative of the subordinate companies, please refer to the above-mentioned duties of the top management representative and HSE management representative.

C Resources

The establishment and operation of the HSE system and the implementation of various activities are inseparable from the support of resources. The senior management should allocate the required resources to the lower departments according to the level. First, based on a comprehensive understanding and mastery of the existing facilities and equipment, the parts that need to be improved and supplemented should be comprehensively considered. When the existing resources cannot meet the requirements, they can be determined in the following order: Facilities that do not meet the requirements of national and local regulations and equipment; facilities and equipment that can meet the company's health, safety and environmental objectives and performance criteria; facilities and equipment required by the company to continuously improve its HSE management level. When allocating resources, you should also consider do what you can, step by step. Under the existing resource conditions, fully consider the requirements of laws and regulations, the obligations, and long-term interests of the enterprise, improve the HSE management system of the enterprise in stages, and give priority to solving key problems. We should neither make unrealistic promises nor set unrealistic goals regardless of financial and material resources, nor should we use insufficient resources as an excuse to weaken HSE management. Conduct economic risk assessments to determine the best use of resources. Solicit the opinions of managers at all levels and experts in various fields. When there is a lack of funds and technologies, the following approaches can be considered: cooperate with users or suppliers to meet common interests; share relevant technologies and experience with peer companies; cooperate with scientific research institutes and universities for cooperative development; Obtain the help of the superior department, and obtain the support of the finance, personnel labor and management and various professional departments. The management should pay attention to the important role of resources in HSE work and conduct regular inspections and reviews on the allocation of resources to ensure that resources are in place and used effectively.

D Capability

Those who pass the assessment can be issued a certificate of employment and start work. Reviews should be conducted every certain number of years. Those who are not qualified in the assessment shall be transferred from the job position, or further training shall be arranged to make them meet the requirements.

Capability mainly refers to the personal qualities of managers, employees and HSE personnel at all levels to do their jobs well, as well as the ability to improve skills and constantly update knowledge through practice. Personnel at all levels, especially those in key HSE positions, should have the necessary capabilities.

People are the key to doing a good job. Many statistical results show that many accidents are caused by poor management or improper operation of key personnel. Selecting qualified managers and operators is one of the keys to avoid accidents. A competency assessment is required to determine whether a person is competent for the job they are doing. Competency assessment should have a set of procedures for personnel with different responsibilities (e. g. , Managers, staff, HSE personnel, key position staff, etc.) . These procedures are different from each other and mainly include, but are not limited to, the following contents: qualifications, referring to education, length of service, etc. ; work performance, including sense of responsibility, work attitude, work performance, etc. ; theoretical assessment and operational assessment, including assessment methods, comprehensive assessment methods etc. ; job training requirements; opinions from all sides For staff involved in key parts of the HSE, the ability assessment must be conducted through formal procedures. These personnel mainly include personnel who play a key role in emergency response, such as managers, commanders and on-site emergency response personnel. Operators engaged in key equipment, such as equipment containing flammable and explosive substances. Equipment with pressure or high temperature. Platform crane operation and maintenance personnel. Personnel responsible for well control and drilling safety and operation. Personnel responsible for environmental pollution control. Radio operators. Supervisory and inspection personnel, including health inspection and environmental monitoring personnel. Personnel who work with the environment in detail.

The successful implementation of HSE management system depends to a large extent on the overall quality and ability of the company's personnel, and the improvement of personal quality and ability mainly depends on education and practice. To ensure that personnel at all levels have high quality and ability, training needs to be done according to specific conditions, so the company should establish a training program. Procedures include training plans, training records, training inspections and improvements.

Training plans are divided into personal training plans and job training plans. The training forms can be divided into classroom lectures and on-site drills. The plan should include the training form, specific steps, and implementation time. Individual HSE training, training and drill plans are prepared to ensure that personnel in key positions or positions achieve, maintain, and update the required knowledge, experience and skills, and an annual plan should usually be developed. The training form, content and time shall be determined according to the basic requirements of HSE and the level and job position of the training object. Key personnel working on HSE must also be trained. The training can be done alone or in combination with other trainings. The annual training plan and various training results should be documented, as should changes in experience and training methods. Classroom assessments, examinations, and training test results should be collated, assessed, and comprehensively analyzed, and the levels achieved by all individuals should be documented. Courses, training, and coaching are evaluated and, if valuable, should be fed back into the design and content of the training program. Experience from training and accidents can also be fed back into the training process if it is believed to improve the criteria and quality of HSE training. The training plan should be updated year by year. Requests for changes to the plan should be submitted to training management for approval and implementation. " Training" is also an important part of the review and audit.

E Contractor

In modern enterprise management, contracting and anti-contracting of some businesses have become common economic activities. To safeguard the interests of the enterprise and maintain a good image of the enterprise, it has become an international practice to require contractors to operate in accordance with their own HSE management models. The contractor's HSE performance is one of the important factors to be considered in the evaluation of bidding and bid qualification. Companies with poor HSE performance will lose the qualification to bid. The contract signed with the contractor also includes the content of HSE management, and the HSE management regulations in the contract will serve as the basis for the contractor to perform HSE management and the basis for Party A to conduct inspections. Enterprises should have documented qualification review procedures. Regarding the HSE management system used by the contractor, one is to use Party A's management system, and the other is to use the contractor's own management system. In the former case, Party A shall provide consultants and inspection and supervision; in the latter case, the contractor shall conduct inspection, supervision, and audit by itself to show Party A that the HSE management system is operating effectively, and Party A shall conduct necessary inspections and external audits, confirming that HSE management has played its due role. The inspection and supervision contents include preparations before the relocation (such as the HSE management system, the integrity of equipment and facilities), the operation process after the start of construction, and the restoration work after completion, etc. Figure 6-19 shows the block diagram of the role of HSE management in contractor management.

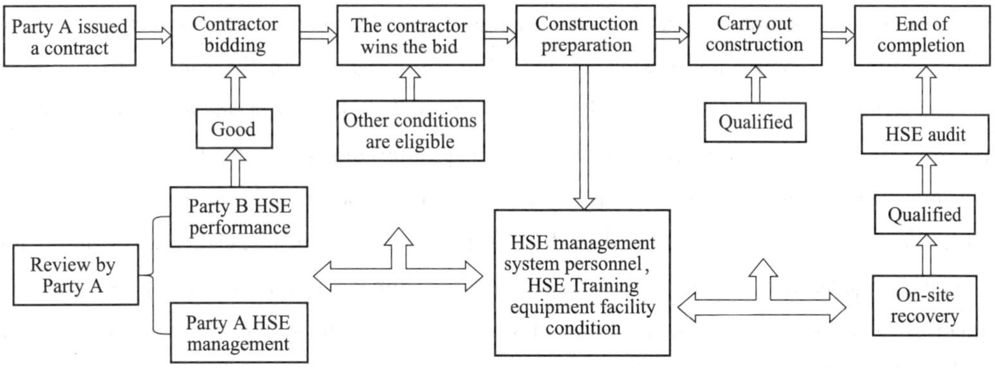

Figure 6-19 Block diagram of the role of HSE management in contractor management

F Information exchange

Information exchange can be divided into two categories: internal information exchange and external information exchange. Internal information exchange is the exchange of information between various levels within the company, such as daily contact between various departments, instructions, regular reports, various messages, reports, writing between various departments, etc. , to coordinate and promote the company's HSE management. It plays a role in guaranteeing the access and distribution of the company's internal information. External information exchange is mainly information communication with all relevant parties (relevant government agencies, groups, the local community, and community public, contractors, customers, etc.) role of image.

For a modern enterprise, in today's informatization, whether it is internal information exchange or external information exchange, it is very important. The exchange of information ensures that the company, contractors, partners and employees at all levels are aware of: the importance of conforming to HSE policies and objectives, as well as their respective roles and responsibilities in the process of achieving these objectives and policies; work activities health, safety and environmental risks and hazards, established prevention, control and mitigation measures and emergency response procedures; potential consequences of violations of approved operating procedures; avenues for recommending improvements to existing procedures to management; the company should formulate a set of procedures for the means of external communication and special emergency arrangements, and specify the methods and channels of information exchange according to the nature of various internal and external information; information recording, processing, reporting objects, disposal methods, etc. For external communications involving important matters, procedures shall be established to specify such matters as the receipt, reply, processing, and filing of information. There are a few things to keep in mind when exchanging information:

In the event of an emergency, the smooth flow of external communication is of special significance, so there should be reliable external communication means and frequent inspections to keep it in good condition. Special measures should also be prepared for special unexpected situations; the company should establish and maintain a set of procedures for exchanging HSE information with relevant government parties to ensure that the company's current policies comply with government laws and regulations; the company should keep secrets under the premise, enabling employees, contractors, customers and companies engaged in the same activities to gain their own HSE experience to help improve everyone's HSE performance; for employees, contractors, customers, government agencies and the public concerned about the company's HSE performance and management A set of acceptance and response procedures should be established and maintained, and a special person should be responsible for it; in the exchange of information, appropriate language and methods should be used at all times, especially when passing technical information to non-professionals. Key procedures and instructions, such as oil spill recovery methods and emergency evacuation instructions, should be in language and methods familiar to on-site staff; means of communication should be regularly tested, and special emergency communication means should be available.

Meetings are an effective means of information exchange, and it is necessary to establish open meeting exchange systems at different levels to conduct top-to-bottom and bottom-to-top information exchanges. The junior manager should chair an HSE leadership group meeting or similar. The meeting plan should include the number, time, and main content of information exchange for HSE meetings at all levels. All employees and contractors should attend relevant HSE meetings, including pre-shift meetings on the job site.

The main purpose and content of the HSE meeting are: to stimulate the two-way and effective communication between management and employees on HSE matters; to promote the participation of all employees in HSE management; to act as a consultant to the management; Implementation effect of measures; organize inspections and reviews for unsafe practices or environmentally unacceptable practices; review inspection and review reports; review investigation reports of recent accidents; review the handling of incidents or accidents that have occurred Inspection; put forward suggestions for further improvement of HSE to the management; strengthen further cooperation of all personnel in promoting HSE; make suggestions for HSE training and guidance of workers.

There are various ways of information exchange to choose from. In addition to meetings held as planned, commonly used information exchange methods for HSE matters are various short meetings, such as pre-shift HSE short meetings; departmental HSE plans and performance target documents; video, Newsletters; Blackboard Newspapers, Post Charts; E-Mails, HSE News, Bulletins; HSE Information Phones; External Safety, Health, and Environmental Data Reports.

In HSE management, information feedback is an important feature of HSE management. The significance of information feedback is information feedback is necessary for review and is the only way to evaluate the effectiveness of information exchange; HSE communication is bidirectional at all levels. the flow of information ceases; respond immediately to critical issues that require prompt and affirmative responses; encourage ideas and suggestions, whether constructive or not; view bottom-up criticism as healthy and normal, should feel uncomfortable in the absence of criticism (indicating that employees may have been reluctant to communicate with management), should abandon the mindset that criticism of upper management is "bad news" and replace it with all HSE criticism is good; consider establishing a system for rewarding feedback information, as a form of recognition, recognition is usually more valuable than material rewards.

G Documentation and control

The primary purpose of documentation is to providea complete description and documented specific operating procedures or guidelines for the HSE management system, and to form a permanent record of the maintenance and implementation of the system. Documents should record accurate information on the HSE management process.

Document control includes management, storage, and circulation, etc. Document control should not only ensure security, but also facilitate access.

The basic framework level of HSE management system document design is shown in Figure 6-20, in which the project HSE plan can be determined according to the specific situation whether it needs to be properad's files are usually divided into the following categories.

Figure 6-20 HSE management system management manual procedure document; management work document; project HSE plan: HSE work instruction, HSE work plan, HSE checklist: other documents. The company shall maintain the following control documents: HSE policy, objectives and planning documents; documents describing the elements of the HSE management system and

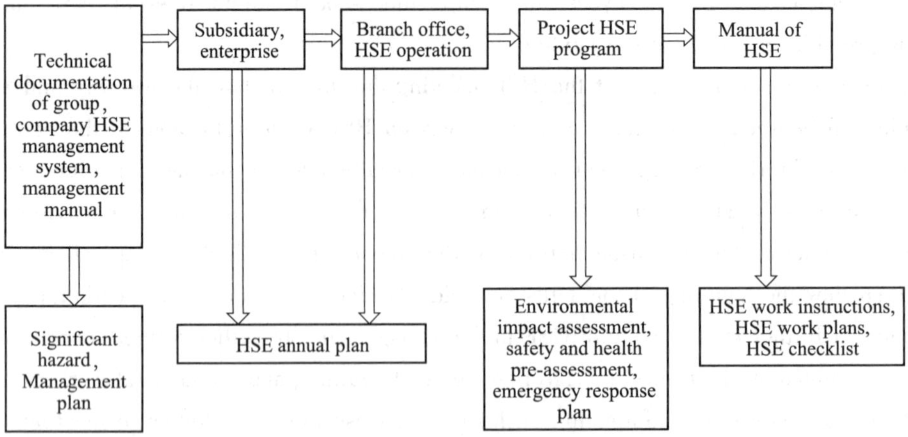

Figure 6-20 The basic framework level of HSE management system document design

their interrelationships; documents describing the interrelationship of documents and their interrelationships with other aspects of the overall management system; documents documenting HSE hazard assessment, management and Documents of risk management results; documents recording relevant legal and regulatory requirements; procedures and guidance documents for key activities and tasks; documents describing emergency plans and responsibilities, and documents for accident and potential emergency response measures; documenting accident investigation and handling processes. Control documents should also include company profile; organizational structure and business units; independent functional and operational units (e. g. , plant design, exploration, land acquisition, drilling, etc.).

Document control is the management of the formed documents. The company should formulate and maintain a set of document management procedures, storage procedures and circulation procedures. This procedure ensures that: documents are marked according to the company, department, function, or activity concerned; documents are regularly reviewed, revised if necessary, and approved for applicability by authorizers before publication; anywhere is available. After the document becomes invalid, it should be immediately removed from all distribution and use places. All documents should be clearly written, dated (including revision date), numbered (including version number), and controlled requirements. It should be clearly marked, kept in an orderly manner, and be properly preserved within a specified period. For document revisions, specific methods and responsibilities should be specified for how to make documents available to employees, contractors, government agencies, and the public. Document control work well can improve HSE management efficiency, can effectively divert information to where employees need and use it, increase responsibility awareness, improve work performance, avoid one-sided information, and show the role of the management system. When determining the level of detail of the HSE management system documents, it is necessary to consider both fullness and refinement according to the specific situation and actual needs of the enterprise, and it is best to obtain the help of system review experts.

The main principles of HSE document control are marked by departments, functional bodies; approved by authorizers before release; clear distinction between what must be followed and what is only for guidance; regular and systematic review, when necessary, or when it is necessary to respond to special requirements, the control procedures shall be modified. Current editions are readily available wherever needed; procedures for changes and modifications are clearly defined; documents are removed from distribution and use immediately when they are obsolete.

6.6.3.4 Evaluation and Risk Management

Preventing accidents and reducing hazards and impacts to an acceptable minimum is the most direct purpose of HSE management system operation. Correct and scientific identification, evaluation and effective management of risks are the key to achieving this purpose. Risk management is an ongoing process that underlies all HSE elements and should regularly check for the presence of hazards and assess the associated risks in business activities. All risks will be managed with appropriate measures to prevent or reduce the impact of a potential accident. China National Petroleum Corporation HSE management system requirements: Hazards and impacts can be identified and prevented, HSE risks of any activities and facilities should be evaluated, action plans and measures to eliminate or reduce risks should be established, and relevant management and control program. Secondary elements of "Assessment and Risk Management" is shown in Table 6-27. And the basic process diagram for evalution and risk managemnt is shown in Figure 6-21.

Table 6-27 Secondary elements of "Assessment and Risk Management"

Secondary elements	Gist
Hazard and impact determination	Systematically identify hazards and their impact on the organization
Establish criteria for judgment	Establish criteria for determining hazards and their impacts
Evaluation	Evaluate identified hazards and impacts based on discriminant criteria
Create documentation describing hazards and impacts	Identified significant hazards and impacts, reduction measures and associated system documents and procedures
Specific objectives and performance guidelines	Establish appropriate and specific harm and impact reduction targets and performance criteria
Risk reduction measures	Select, evaluate and implement measures to reduce risks and their impacts

Identify—What could be wrong?

Evaluation—The nature of the problem? Risk?

Control—Is there a better way to control it? Is the control method adequate?

Review—Is it implemented? Can adverse consequences be controlled? Is the review sufficient?

It is divided into four stages to facilitate the description of the entire risk assessment and management process, but in fact the boundaries of these stages are not always clear. decision making. That is, after determining and evaluating the risk control process, if the control process is considered insufficient, the criterion should be reconsidered, and the risk evaluation should be re-

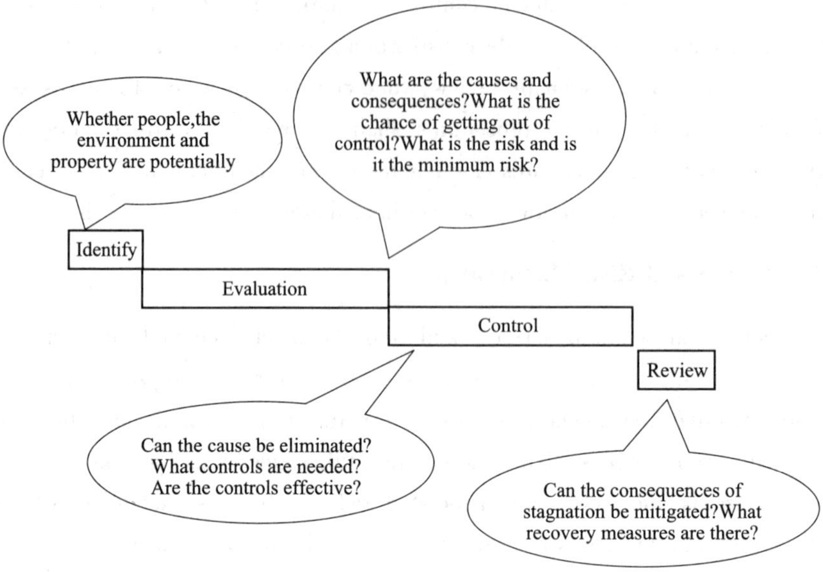

Figure 6-21 Basic process diagram for evaluation and risk management

performed to determine further risk control measures. In more complex cases, these processes may be repeated. However, in any case, the result of risk identification and evaluation should be such that the hazard is reduced to "reasonably practicable and as low as possible". The evaluation and management of risks will be introduced in detail in the following chapters.

6.6.3.5 Planning

Planning is an important part of implementing HSE risk management and an important aspect of implementing HSE plan management, see Table 6-28. CNPC's HSE management system requirements: HSE planning is part of the company's overall planning and should focus on HSE goals and performance criteria at different levels. Determine hazard reduction measures through hazard and impact management procedures, implement special funds, necessary equipment, and resources, and form an operational plan.

Table 6-28 Secondary elements of "Planning"

Secondary elements	Gist
General	Develop HSE management plan and set work goals and steps
Integrity of the facility	The design, construction, procurement, operation, maintenance, and inspection of engineering or critical facilities meet established objectives and specified performance criteria
Procedures and work guidelines	Documented procedural documents and guidelines for activities and tasks
Change management	Controls implemented for permanent or temporary changes to people, facilities, processes, procedures, etc.
Emergency response plan	A plan for taking precautionary measures against emergencies

A General

The company shall develop and maintain a plan for the achievement of HSE objectives and performance criteria throughout its own work procedures. The plan shall include: a clear statement of the objectives; clarifying the roles of all levels of the organization in establishing and achieving the objectives and performance criteria; achieving the objectives and Methods of performance criteria; resource requirements; timelines for implementing plans; programs to motivate employees to work towards a suitable HSE culture; mechanisms for providing feedback to all personnel on HSE performance; reward scheme); evaluation and improvement mechanisms.

A viable improvement plan is a key component of the HSE management system as a means of achieving HSE corporate policy and objectives. These plans require adequate resources and public commitment from all staff. Ideally, these plans should be part of the company's overall corporate plan. Activities that require the development of an HSE plan may include new designs; existing operations: retrofitting of existing facilities; decommissioning plans; geological surveys; seismic surveys; exploration and development plans. To ensure adequate manpower and the necessary budget commitments, resource requirements and schedules should also be specified. It is advisable to consult with regulatory experts and other external organizations involved in the development of regulations, rules, and standards to ensure that possible future regulatory and regulatory requirements are considered when planning (wherever possible). Corresponding to the activities and tasks at the three levels of the group company, the operation company and the operation team, there are plans for each level, as shown in Table 6-29. The table also shows the time required for each type of plan and the included plans. The level of detail of the content.

Table 6-29 Plans at three levels (company, operational processes, and tasks) and their levels of detail

Organizational level	Plan	Required time	Level of detail
Group company	Set strategic direction	Several years	Low
Operating company	Management prepares a business plan	1 to 5 years	Middle
Work team	Action plan	Weeks	High

At the group company level is mainly responsible for the formulation of plans related to policies and strategic objectives, and the following two levels correspond to most plans related to operations and projects, respectively. The planning of HSE management implications includes the formulation of strategies to meet the strategic objectives of the entire company, various agencies, and departments and to achieve specific goals, as well as the respective goals formulated by each agency and department to meet the strategic goals. To improve environmental performance, planning at all levels is very important, because reducing wastewater and air emissions is a long-term process that requires pre-budgeting and resource allocation. The HSE management plan at the operational process level mainly consists of the following framework components: organization, roles, responsibilities; information exchange; qualification and training; contractor management; standards; prohibited items: waste management; emission control. Plans at the HSE task level include the following to ensure that tasks are accomplished: adequate human resources; adequate equipment, tools, and services; safe work locations; safe work methods; procedures; reasonable time constraints. The plan at the task level should be detailed enough to clearly state: What needs to be done? Why do it? Who is responsible for doing (Who)? Where? When to do (When)? How? These are identified in the previous hazard assessment as the necessary control requirements for managing the risk. The minimum control requirement is that they must meet the limits required by regulations. But in many practical situations, such regulatory requirements are lacking. During the development phase of the plan, it is desirable to have people involved in implementation, but this is not always possible. Where this is not possible, the key personnel concerned should be familiarized with the plan before the task begins. Even if key personnel are involved in developing the plan, they should be reminded of key HSE matters before executing tasks as planned.

B Integrity of the facility

"Facility" refers to the facilities related to health, safety, and environmental protection; "Integrity of the facility" refers to the facilities related to health, safety, and environmental protection, which exist at the same time as the main facility and are in good operating condition. The health, safety and environmental protection facilities shall be designed, constructed, and put into operation at the same time as the main facilities, and the operating conditions shall meet the specified requirements. The company shall establish and maintain facility integrity procedures to ensure that the design, manufacture, procurement, use, maintenance, and inspection of HSE critical facilities and equipment are both fit for purpose and in compliance with prescribed guidelines. The pre-purchase and pre-construction evaluation of new facilities and equipment should include a detailed [if performance evaluation] to meet HSE requirements. It should be emphasized that the design phase is the best time to reduce risks and prevent adverse impacts. The procedures and systems to ensure the integrity of the facility shall, among other factors, emphasize the integrity of the structure and the effectiveness of the process control, ignition control and protection, detection, shutdown, emergency response and life-saving systems. Deviations between the design and the standard shall only be allowed after review and approval by authorized personnel or supervisors, and the reasons for the deviation shall be documented. For critical equipment and related activities, it is also important to follow recognized codes of practice and standards. To contribute to the integrity of existing and planned facilities, the integrity of the process, especially the design, manufacture, installation, maintenance, testing, and inspection of critical equipment, should be carefully reviewed. Here, critical equipment refers to the equipment identified during the evaluation process that is critical to continued effective HSE control. Emphasis should be placed on the design of new facilities, where early hazard identification allows us to take the best risk reduction measures and eliminate hazards at source to prevent accidents. Quality assurance measures in the manufacture of critical equipment help ensure that materials and manufacturing processes meet design standards. The installation process must be managed and checked to ensure compliance with design standards and manufacturer's requirements. Reviews of maintenance, testing, and inspection system effectiveness help ensure the continued integrity of critical equipment. All personnel performing the above tasks in relation to the integrity of the facility are HSE key personnel and therefore should have the necessary experience, qualifications, and training to ensure that they can undertake these important risk reduction measures. To ensure that the integrity of the facility is protected, the following should be done first: clearly define the responsibilities of facility management; practical and comprehensive understanding of the planned change control system; development of open inspection principles and procedures; development of documented management and cross-checks Principles (Mutual Inspection of Facilities) Procedures. An analytical technique shall be established or established to identify the following activities and facilities: critical activities and facilities dedicated to health, safety, and environmental protection, such as flares, sewage treatment facilities; Critical activities and facilities, such as equipment on equipment for fire protection, emergency shutdown, alarm, and noise reduction; other critical activities and facilities.

C Procedures and work instructions

"Procedure" means the prescribed means for carrying out an activity. In many cases, procedures can be documented and are referred to as "written procedures" or "documented procedures". The procedures usually include: the purpose and scope of the activity, what to do and who to do it, when where and how to do it, what equipment, materials and documents should be used, how to control and record the activity, etc. A "job guide" sets out the procedures for completing tasks on the job site and is usually provided to the performer of the task in the form of a manual.

Development of procedure documentscan be called HSE key tasks must be developed procedures, the so-called key tasks refer to activities that include one of the four elements (identification, evaluation, control, and review) of the hazard and impact management process. Activities that violate HSE policy, regulatory requirements, or performance standards due to the lack of documented work procedures should be identified and documented procedures and guidelines established for these activities. It should be specified in the document how the task will be carried out, whether it is carried out by the company's own employees or by the entrusting party, to ensure the integrity of the technology and the effective transfer of knowledge. As far as possible, the personnel responsible for implementing these procedures and guidance documents should be involved in the development of the documents. It should strive to write these documents in a concise format and language and cover as precisely as possible the activities they involve. All documented work procedures should be the presentation is simple, clear, and easy to understand. It should clarify the responsibilities of personnel, the methods used, and, if necessary, the performance standards and criteria to be met. Monitoring requirements and the need for personal protective equipment can be specified in the work instructions. In a word, when formulating documents, it should be noted that the preparation should be well planned according to the company's situation and risk identification results and strive to reduce the number and length of documents as much as possible under the premise of achieving control; there is a necessary connection between procedural documents. But avoid excessive repetition. Take advantage of the applicable procedures your company has in place. The document is actionable: the content is simple, clear, easy to understand, and individual responsibilities are clear. Purchasing and contracting should develop procedures to ensure that suppliers and company principals behave in accordance with a company policy relevant to them. Guidance on field work tasks can be provided in a variety of ways, depending on the complexity of the task, the capabilities of the person performing the task, the inherent hazards and risks associated with it, and the impact on other aspects of the facility and operations. For example, production facilities that store or produce hydrocarbons have potential fire and explosion hazards and require strict controls and management, so most work should be carried out under the control of a permit-to-work system. Here, the guidelines identify what to do, precautions and how to notify other parties who may be affected, and require all parties involved to post a work permit sign.

The work guidelines specify how tasks are to be done on the job site, whether the work is carried out by the company's own employees or by an entrusting party. For HSE-critical tasks, if not performed correctly, there is a high likelihood of adverse HSE consequences. Therefore, these work instructions should be documented and communicated to relevant personnel. For all personnel, including contractors, work instructions for safe operation, maintenance, and renewal activities, including coordinated operations, shall be provided. For example, the following operations often require specific safe work guidelines: opening of equipment and piping; cutting and interlocking of electrical and mechanical equipment; high temperature operations, including other ignition operations; access to exits in confined spaces crane operation; working at heights; Pipe-laying operations affecting other equipment or other offshore installations; transport of hydrocarbons through offshore shipping facilities; use of explosives; use of radioactive sources; operations affecting the integrity of safety systems; operations in flammable gas vent areas. For tasks involving operations with, a work authorization (work permit) system and a strict shift system should be established to ensure the smooth progress of work activities. After the work guide is formulated according to the prescribed procedures, it is formed into a document, signed by the authorized person, and issued for implementation, and notified to the relevant units and personnel.

D Change management

For temporary or permanent changes to people, plants, processes and procedures, the company shall establish and maintain planned management procedures to avoid adverse effects on HSE. Any changes in the company's personnel, equipment, processes, and procedures have the potential to adversely affect HSE. All changes should be considered, not only those mentioned above, but also those brought about by restructuring of the organization, such as those brought about by acquisitions, mergers, new joint developments, and additions of partners. Change-related plans must address HSE issues at all stages affected by the change to ensure that hazards or environmental impacts are minimized through effective planning and design. Examples of changes that may affect the entire HSE program are listed below.

a Changes to facilities

Changes to facilities may be caused by changes in process or mechanical design, as may changes in production media, additives, product specifications, by-products, scrap, design terms, monitoring instrumentation and control systems, or building materials. Facility changes may occur under the following conditions: construction of a production or process facility. The construction of new facilities, including the supporting connection of production or processes to existing facilities, the reconfiguration of facilities or the modification of existing facilities (equipment). Modification of an existing facility resulting in a change in the design and structural support, layout, configuration of the facility or equipment. Projects that increase facility throughput or accommodate different media. Significant changes in operating conditions, including changes in pressure, temperature, flow rate, or other process conditions from the original process or design. Equipment changes, including adding new equipment or retrofitting existing equipment, are mainly changes to alarm, detection, and control procedures. Changes to the process or equipment can result in changes in facility depressurization or venting requirements, including throughput, increased operating temperature or pressure, expanded equipment specifications, or additional equipment requiring higher pressures. Jumper of equipment in normal use. Work outside the scope of existing operating procedures, including start-up, normal shutdown, and emergency shutdown. Changes in mechanical design involved in the risk assessment and management of various elements of the HSE management system. The introduction of new or different chemical agents (e. g. , Preservatives, anti-scaling agents and anti-foaming agents). Changes to facilities include changes to machinery (such as drilling equipment, construction equipment, temporary connections, or replacement of failed components), some of which may not be noted in the process and testing flow charts. These may include Replacement with specifications different from Machinery and equipment required to be considered in risk assessment and management; Temporary pipelines, connections, hoses, or repaired pipelines; Alternative supplies of process materials, catalysts, or reactants; Connection of temporary electrical equipment or utility power, Excludes emergencies. Program changes related to process control or safety systems.

b Change to personnel

Changes to personnel include changes in contractors and changes from the organization, changes in equipment and changes in operators; changes in organization due to facility transfer require the company to revise the HSE management system.

c Changes to procedures

Changesto the operational process and corresponding procedures involved in the risk assessment of the HSE management system and each element of the management system.

d Changes to laws and regulations

The company studies the content of the promulgated or newly promulgated laws and regulations to make the HSE management system conform to the requirements of the laws and regulations. Depending on the nature of the change and its potential impact, change management procedures should include, in addition to applying to the HSE issues involved, the identification and documentation of proposed changes and their implementation. Responsibility for reviewing and documenting the HSE hazards that may arise from changes and their implementation. Documentation of approved changes and their implementation procedures, including methods for identifying HSE hazards and evaluating, reducing risks and impacts; information exchange and training requirements; time limits if necessary; verification and monitoring requirements; measures. The authority or authority that approved the implementation of the change. For possible changes of HSE key matters, firstly obtain the document of change description or plan, review before implementation, and make necessary supplements to the HSE management system, if necessary, to ensure that the introduction of changes will not adversely affect the good HSE performance. The change management procedures should be explained as to how the company explains and evaluates the new or supplementary regulations to be implemented, and how to integrate the revised rules into the HSE management system. When significant changes in HSE are caused by new operations (such as those related to purchase, establishment, obsolescence, and products, services or processes) or modified operations, a dedicated change management plan should be established to determine: HSE objectives to be achieved: HSE objectives to be achieved methods; resource requirements to achieve HSE objectives; procedures for dealing with changes or changes in the progress of the project; adequacy of corrective and monitoring methods employed, and how they will be implemented. To facilitate the implementation of change management, standards and guidelines should be clearly divided into mandatory and directive requirements. Gradual changes in emissions composition or production beyond the original design range should be given special attention and vigilance in change management. Such changes are slow and often overlooked but are likely to change from quantitative changes to qualitative changes, which may lead to accidents. Likewise, changes such as changes in the chemicals used or the encroachment of land adjacent to processing plants by local dwellings may lead to changes in the environment that need to be re-evaluated.

e Emergency response plan

The company should take preventive measures against potential emergencies and accidents, and formulate emergency response plans, so that emergencies and accidents can be dealt with quickly, promptly, and effectively, to ensure that possible losses are minimized.

Emergency responses are developed through a risk analysis. The establishment of the HSE management system by the company reflects the idea of prevention. Although the occurrence of accidents can be prevented through the management system, it does not necessarily mean that accidents will not occur after the establishment of the system. The reasons for accidents may include the following aspects: HSE management system There are imperfections; employees do not understand the requirements of the HSE management system; employees understand the requirements, but do not follow them; accidents occur. Emergencies requiring contingency planning may include fires and explosions; toxic gas releases; failure of critical controls for electrical power or protection; structural failures; on-site injuries; diving, marine and aviation accidents; spills and fugitive discharges of other substances; loss of radioactive material; breach or intentional breach of confidentiality; disease outbreaks; geophysical or natural events.

The company shall establish and maintain procedures for identifying foreseeable emergencies through systematic review and analysis, namely emergency response plans. The identified possible risks shall be recorded and updated regularly to ensure an effective response toemergencies. The emergency response plan includes: organization, responsibilities, authorities, and procedures for emergency response and disaster control, including internal and external communications; systems and procedures for providing personnel avoidance, evacuation, rescue, and medical treatment; prevention, mitigation, and monitoring of emergency response Systems and procedures for environmental impacts of actions; procedures for communicating with authorized persons, relevant personnel and interested parties; systems and procedures for mobilizing company equipment, facilities and personnel; arrangements and procedures for mobilizing second-party resources for emergency support; training Arrangements for emergency response teams and testing emergency response systems and procedures. The company develops an emergency response plan, which should be communicated to command and control personnel; emergency services; employees and contractors who may be affected; others who may be affected. Among other factors, to evaluate the effectiveness of the response plan, the company should establish procedures for regular or non-periodic verification of the response plan, which can be carried out through field exercises or other suitable methods. Based on the experience gained, the emergency response plan can be modified as necessary. Procedures should also be established to periodically review emergency equipment requirements and maintain equipment in standby. In summary, an emergency response plan needs to be clearly communicated; carefully rehearsed; cooperating with internal and external emergency response teams; paying particular attention to external communications; having a mindset to report and investigate incidents, such as the effects of an out-of-control fire spilling with water) to control. The emergency system is usually divided into three levels, namely the first line, the second line and the third line, and all three levels have their own emergency measures. The first line is the unit that is directly endangered, it is the place that directly fights against the harm and needs rescue, usually the work site; the second line is the upper management department of the first line, responsible for production management and supporting the emergency rescue of the first line, such as contacting the first line. There is no rescue equipment (helicopter, etc.); the third line is the headquarters of the group company, in addition to being responsible for strategic management and issuing important instructions, it is mainly responsible for reporting major events to the government and releasing information to the outside world. From top to bottom, there is a reliable and convenient information transmission system to ensure the effective implementation of emergency response. Analyze all risks and establish corresponding emergency measures for each risk, including specific steps and specific instructions for each step. For example, if a person falls into the sea, the stipulated rescue time is 7 minutes, and the equipment used, participants and specific action steps are stipulated, and prior training has been carried out.

6.6.3.6 Implementation and Monitoring

Implementation and monitoring are key to the implementation of the HSE management system. China National Petroleum Corporation HSE management system requirements: When employees and contractors start to contact any work, they must be familiar with the relevant HSE control measures and implement the work according to the procedures established in the planning stage, operation guidelines and relevant HSE policies, and monitor it.

This element contains 6 secondary elements, see Table 6-30.

Table 6-30 Secondary elements of "Implementation and Monitoring"

Secondary elements	Gist
Activities and tasks	Activities and tasks carried out in accordance with work instructions or procedures
Monitor	Monitor HSE performance, establish and keep corresponding records
Record	System of record
Non-conformance and correction	Determination of non-conformities and corrective actions for non-conformities
Accident report	Document and report incidents that have affected or are affecting HSE performance
Accident investigation and handling	Investigation of accidents caused by various situations

A Activities and tasks

In the process of HSE management, personnel at different levels of the company have different responsibilities and assume different tasks. Personnel at different levels should carry out and execute their respective activities and tasks in accordance with the work procedures and guidelines at the planning stage or earlier in accordance with the HSE policy, for example: managers and management: follow the HSE policy to formulate strategic goals and high-level activity plans. Management executive level: Develop written instructions (usually multiple tasks) related to activities in the form of plans and work procedures to guide each work. Operational level: Here, plans and work procedures are written and issued in accordance with established safe work systems (e. g. , Work permits, cooperating procedures, start-up procedures, permitting operation manuals), and workers follow these regulations. Required to complete the task. Management is responsible for ensuring that activities and inspections are carried out in accordance with the relevant procedures. Commitment to implementing policies and plans includes, among other responsibilities, ensuring that tasks performed meet HSE objectives, do not violate performance criteria, and do not exceed control limits. Managers should ensure that the company's HSE performance is consistently in good shape through continuous monitoring. To ensure the effective implementation of activities and mission plans, all levels of the company must comply with work procedures and guidelines, and the company and contractor employees should be familiar with the relevant work procedures and guidelines before starting work.

B Monitoring

The company shall establish and maintain working procedures, monitor the relevant situation of HSE performance, record and save the corresponding monitoring results, to facilitate the continuous improvement of HSE performance. For the relevant activities and areas, the company shall perform the following work: check and record the monitoring information obtained, and specify the required accuracy of the monitoring results; specify and record the monitoring method, monitoring location and measurement frequency; establish, record and maintain the control measurement methods for quality; establishing and documenting data processing and interpretation methods; establishing and documenting actions to be taken when monitoring results exceed performance criteria; evaluating and documenting the validity of affected data in the event of a monitoring system failure; protecting monitoring systems from damage and Unauthorized verification.

Both routine monitoring and emergency monitoring require work procedures. Routine monitoring provides information on normal conditions, accident conditions, such as health problems and environmental damage; checks(e. g., work procedures) for compliance with HSE requirements and compliance with objectives and performance criteria. Emergency monitoring provides information about an accident that has occurred (including accident hazards, health problems, and environmental damage), and from this leads to measures to prevent similar accidents in the future. Monitoring provides a means of measuring HSE performance and may include activities such as: regularly reviewing progress in implementing the HSE plan against overall and specific objectives; regularly inspecting facilities, plants and equipment against prescribed performance criteria; systematically observing the work of front-line supervisors and performance, and assess compliance with work procedures and guidelines; regular monitoring of waste water, air emissions and waste disposal; employee health checks, including exposure measurements and medical examinations. Monitoring the control of HSE key activities and processes, the content and frequency of monitoring must reflect the nature and extent of the risks involved and the concentrations in the areas where the greatest benefits are generated: "high risk" facilities, plants, activities and tasks require more detailed, higher frequency monitoring.

C Records

Records are part of the HSE management system documentation and should be documented at all important links in the management process. It is listed here as a separate item due to the extensive documentation support required during implementation and monitoring. To illustrate and compare the status with the HSE policy and its requirements, and to document the degree to which program objectives and performance criteria have been met. The company shall establish a set of procedures to ensure the completeness, accessibility, and control of records: related contractor and selection records; audit and review results; training records; employee medical records; safety records; environmental monitoring records, etc. Record retention periods and registration systems should be specified, and procedures should be established to ensure the availability and confidentiality of records. Records are evidence that the HSE management system is operating, therefore, care should be taken to focus on the records necessary to implement the management system and on the degree to which objectives, indicators and tasks have been achieved. Records already formed in other parts of the overall management system need not be repeated, but the method of how to obtain them should be specified. Excluding legal and regulatory requirements and records of significant hazards and environmental impacts, records shall include audit and review reports; status and improvement actions not in compliance with HSE policy; all incidents and follow-up actions; information on appropriate suppliers and contractors; inspection and maintenance reports; product identification and composition data; monitoring data; training records. Records provide historical information on reportable incidents and individual instances of non-compliance with the HSE management system, thus providing useful information on long-term trends. For example, analysis of medical treatment records may show increasing complaints of stomach pain, indicating possible problems in food processing areas or drinking water systems; similarly, analysis of oil in produced water may show increasing trends over time, indicating reservoir fluids. There may be changes, or maintenance or modifications to the treatment system may be required.

D Non-conformities and corrective actions

"Non-conformity" is a term used to refer to various deviations or violations of HSE policies, objectives, targets, or any other system requirements. The company shall determine the responsibilities and persons in charge of implementing corrective actions when a non-conformity with the prescribed requirements of the management system and its operation and consequences occurs. Non-compliance can be determined through monitoring procedures, through information from employees, contractors, customers, government representatives or the public, or through accident investigations. The company shall establish and maintain procedures for conducting investigations and implementing corrective actions through which individual actions or related activities are managed under the leadership of a management representative. The management shall: notify interested parties; determine cause or root cause; develop action plan or improve the plan; implement preventive measures that match nonconformities; conduct control management to ensure that all implemented preventive measures are effective; revise procedures, strengthen measures to prevent the recurrence of accidents, and notify relevant personnel of changes to implement. Accidents caused by non-compliance with regulatory requirements may be sudden, temporary, or may last for a long time. Causes may be defects or deficiencies in the management system itself, or problems in factories and facilities, or from human error. When conducting a non-conformity investigation, the accident cause mechanism, including internal factors of the management system, shall be fully established, and reported. Such investigations can assist in the development of an improvement action plan, including measures applicable to return to "Compliant" status as quickly as possible; prevent recurrence; evaluate and mitigate all adverse HSE impacts, such as the mutual good support of the quality management system; evaluate the effectiveness of the above measures. The implementation of improvement measures cannot be considered complete until the effectiveness of the above has been demonstrated and appropriate modifications to procedures, documents and records have been made. When improvement actions involve work procedures that take a long time to complete, the improvement actions should be made part of the management plan.

E Incident reporting

Organizations at all levels should establish and maintain procedures for recording and reporting internal incidents that have impacted and may be impacting HSE performance to learn lessons and take necessary action. Accident investigations should begin as soon as possible and should consider the needs of the accident site, personnel, and environmental protection. There should be a clear mechanism for reporting incidents to law enforcement. Incident reporting should be to the extent required by law, or more broadly as required by the company's external communications policy. It is very important for employees to report all accidents to learn lessons and improve the HSE management system. A reporting system should be established to encourage employees to report rather than blame. The key elements required to be reported by the reporting system include details of all injuries, occupational diseases, or adverse environmental effects; number of people involved in the injury; description of environmental conditions; details of the incident; details of the results; possible consequences; adverse effects caused by. Injury to persons and damage to property usually require immediate and comprehensive reporting, but minor incidents that do not result in injury or property damage (called "incidents") are more frequent, and the same incidents are caused by slightly different causes. The environment may cause a major accident. But often such informative signs of incidents go unreported, either because their potential importance is not felt, or they are afraid to report, for example, for fear of blame, or because reporting systems are too complex. Therefore, the reporting system should be simple to encourage reporting of incident incidents and to identify potentially higher risk incident incidents to obtain critical information. The main principles of accident reporting include statistical volume: The operating company's own use of data usually needs to be recorded in more volume than the group company's annual report needs. Coverage: statistics should cover HSE key operations and all employees, as well as include separate statistics for contractors. Classification: the report should be classified as required. Standardization: reports should use a standard format. Behavioral indicators: safety is reported with four types of indicators, namely fatality frequency, lost-time injury frequency, total reportable accident frequency and severity, health is reported with two types of indicators, namely exposure characteristic indicators and occupational disease characteristic indicators, and environmental indicators are mainly used in: organized and unorganized air emissions to the atmosphere; organized and unorganized wastewater discharges to surface water; organized and unorganized waste disposal; contamination of soil and groundwater; accidents (complaints, fires, leaks, and various compliance); use of energy and natural resources. Whether it is a sudden accident, or an accident caused by a potential defect of the HSE management system, it should be identified so that the personnel responsible for the accident handling can make a judgment. Procedures and responsibilities for handling incidents should be clearly defined, and this procedure should be substantially like the process for implementing corrective actions in the event of a non-conformity with the HSE management system. All accidents, including those with a higher potential hazard, require proper investigation to identify the cause of the accident, determine the actions to be taken to prevent it from recurring, and meet the legal requirements for investigation and reporting; and provide a true record of the conditions in which the accident occurred.

The investigation process consists of the following basic steps: notification, initial review, and accident report; determination of need for further review, designation of an investigation team; investigation itself includes review of accident site and environmental conditions, interviews with witnesses, analysis of operating conditions, data and other evidence; prepare investigative reports and agree on remedial actions.

After the initial review and reporting of the incident, the first step is to determine the appropriate level of investigation, depending on the severity of the incident and its actual and potential consequences. The seriousness of the actual consequences should be clear, but the seriousness of the potential consequences is not necessarily clear. The size and composition of the investigation team will depend on the specific incident. It is important that team members are adequately trained to perform their tasks objectively, fairly, and effectively. The primary role of accident investigation is to determine probable causes and appropriate remedies. Therefore, the investigative team may require support and authorization from company leadership to obtain the necessary information and assurance to implement the identified remedial actions. The investigation and handling of the accident includes two parts: investigation and taking remedial measures. The progress of the implementation of the remedial measures must be monitored, and the remedial actions cannot be considered complete until no effect is shown.

6.6.3.7 Audits and Reviews

Audit and revieware the last link of the HSF management system, details are shown in Table 6-31. It is a regular assessment of the performance, effectiveness, and continuous applicability of the HSE management system, and is a necessary guarantee for the continuous improvement of the system. China National Petroleum Corporation HSE management system requirements: HSE audit and review is the responsibility of company management. All specifications implemented on site and in the production, process should be regularly inspected and audited to evaluate the compliance of HSE management standards and related regulations, suggesting areas for continuous improvement.

Table 6-31 Audit and review'secondary elements

Secondary elements	Gist
Audit	Internal or external audits initiated by the company itself
Review	Regular review of the HSE management system and its performance by the company's senior management

China National Petroleum Corporation's health, safety and environmental management system standard is an equivalent transformation of the International Organization for Standardization ISO/CD 14690 "Petroleum and Gas Industry Health, Safety and Environmental Management System" (draft standard), which has been approved by the world Recognized by several major oil companies, the system elements listed in Table 6-32 should be used as our primary reference for establishing the HSE management system.

Table 6-32 Key elements of HSE management system

Elements of HSE management	Main content
Leadership and commitment	A top-down commitment to building and maintaining an HSE corporate culture
Policy and strategic objectives	Intentions for health, safety and environmental management, principles for action, and goals for improving HSE performance

Continued Table 6-32

Elements of HSE management	Main content
Organization, resource and document management	Personnel organization, resources and complete health, safety and environmental management system documentation
Evaluation and risk management	Determination and evaluation of health, safety and environmental risks in activities, products and services, and development of risk control measures
Planning	The implementation plan of work activities, including the selection of risk reduction measures through a set of risk management procedures, involving the planning of existing operations, the management of changes, and the development and updating of emergency response measures, etc.
Implement and monitor	Execution and monitoring of activities and how corrective action is taken when necessary
Audit and review	Periodic evaluation of system performance and adaptability

However, the establishment of the system is not the ultimate goal, the implementation and operation of the system and the realization of HSE goals are what we require. Therefore, when establishing your own HSE management system, when determining the mode and content of the system, you should also consider the nature, characteristics, and scale of your own enterprise, as well as the original management mode and management system, to establish a suitable for the enterprise, with the original management system. Enterprise-specific HSE management system.

Break the HSE policies and objectives down to the basic units of the enterprise. Implement the measures and responsibilities for hazard identification and risk reduction to the post personnel, so that the HSE management system can operate in a standardized manner from top to bottom. It will embody the work requirements of "all staff participation, risk control, continuous improvement, and performance assurance". For example, CNPC has adopted a complete set of *HSE Operation Instructions*, *HSE Operation Plan* and *HSE Checklist* which is referred as "two documents and one table" to implement the HSE management system. The basis of "two books and one table" is HSE risk management. The key of "two books and one table" is to implement responsibilities. The role of "two books and one table" is to promote continuous improvement.

With a sound safety, health and environmental management system, or have already carried out quality management system certification or environmental management system certification, why establish an HSE management system?

Doing a good jobin HSE management is not only the obligation stipulated by law, the requirement of the government, but also the needs of the society and the vital interests of enterprises. All enterprises that do well in health, safety and environment work will have a set of sound, documented and feasible safety, health and environment management methods and systems. Safety, health and environmental management are done well, the safety and health of employees are guaranteed, the company's property is not lost, and the environment is protected, which can promote the development of productivity and improve economic benefits. At the same time, it can establish a good image of the company and cultivate A group of good employees enhances market competitiveness. The main purposes of HSE management are to meet the government's legal and regulatory requirements for health, safety and the environment; to provide guarantees for the realization of the general policy, general objectives and specific objectives proposed by the enterprise; to reduce accidents and ensure the health and safety of employees. Safety, protect the enterprise. protect the environment and meet the requirements of sustainable development; improve the utilization rate of raw materials and energy, protect natural resources, and increase economic benefits; reduce medical treatment, compensation, property damage costs, and reduce insurance costs; meet public expectations, maintain good public and social relations; maintain the reputation of the enterprise and enhance market competitiveness.

It is difficult for the existing management system to meet the requirements of establishing modern enterprise management. The main manifestation is thatalthough the enterprise has a set of existing and effective management methods and management systems, they are in charge of each side, and health, safety and environmental management sometimes have their own set, which has not been formed. Scientific, systematic and continuous management system still needed to improve. There is a big gap between the thinking mode of health, safety and environmental management and foreign advanced management thinking, such as the general lack of foreign high-level commitment and "zero accident" thinking mode, lack of modern system management method and scientific management mode required by enterprise HSE management.

The oil industry is a high-risk industry. Health, safety, and environmental risks are associated and should be managed at the same time. The health, safety and environmental accidents of oil companies are often interrelated and must be controlled at the same time. ISO quality management system and ISO 14000 environment compassion systems are advanced management systems that also include some health and safety elements, but they are mainly aimed at quality and environment and do not form a whole. The establishment of HSE management system is the need for enterprises to connect with the international market. Almost all large oil and gas companies in the world are implementing this advanced H-energy management model. Good HSE management is a quasi-personal certificate for entering the international market. It can ensure the continuous improvement of the HSE management level and the reputation of the enterprise, and increase the competitiveness in the international market.

With the development of globalization and petroleum economy, health, safety and environmental management systems have been widely implemented in the international petroleum industry. Many large oil and gas companies have gained many useful experiences in the process of establishing and implementing HSE management systems, such as the establishment of corporate culture, health, safety and environmental excellence plans, external verification of HSE performance annual reports, ISO 14000 certified HSE management. The implementation of the system, and the establishment of an integrated quality, health, safety and environment (QHSE) management system, etc.

A External verification of HSE annual report

At present, about 53% of the top100 oil and gas companies in the world publish their HSE annual reports every year. In order to reflect the reliability of the reported data, about one-third of the companies conduct third-party verification of their HSE annual reports. Reasons for implementing third-party verification of HSE reports: to increase transparency. The HSE performance information provided by the report is transparent and enhances mutual trust between companies and shareholders.

Guaranteed integrity. The verification process evaluates the data collection process and the collected data through recognized evaluation procedures, which ensures the completeness and accuracy of the report contents and avoids the suspicion of self-bragging. Promote improvement. The verification process can make a strict evaluation of the HSE data collection, which not only improves the reliability of the data, but also enables the data collection process to be continuously improved.

Bring credibility. Many organizations choose a verifier considering the credibility that the verifier's reputation can bring to the information in the verification report, so verification can be seen as a means of increasing credibility and giving readers more trust in the report.

Validation focuses on the following aspects: reporting accuracy. Verify the reliability of the information in the report, that is, the accuracy of the content and data. Comprehensiveness of the report. Verify that the content of the report regarding overall performance is consistent with the priorities given by the company and those deemed important by the external audit. other aspects. Evaluation beyond the content of the evaluation report, such as evaluation of HSE management process, evaluation of performance improvement, consistency with shareholders' views, etc.

Validation usually consists of the following basic steps.

Site visit: through a site visit survey, to evaluate the reliability of the data and processing methods, the implementation of company guidelines, the inclusion of all exhaust, waste water, and slag sources, and the competence of the technicians Wait.

Data verification: Evaluate the process of data transfer from the site to the company, the process of data inspection and verification, and the communication of existing problems and verification through data verification.

Report verification: By verifying the contents of the report, one evaluates the accuracy of its statements, the suitability of the data, and the conformity of all internal information and priorities.

Validator Statement: Statement that includes an overview of the validation methodology, reporting accuracy, reliability and consistency of data processing, and other relevant aspects.

B Implementation of ISO 14000 certified HSE management system

Due to the continuous improvement of people's environmental awareness and the unique significance of the environmental management system, especially the limitation of " green trade completion" , it seems that the major oil companies in the world are implementing the HSE management system while carrying out ISO 14001 certification. With this trend, many companies have established timelines for achieving certification. Shell requires all companies to commit to the timetable for ISO 14001 certification in their HSE goals in 2000, and will use the ratio of contractors and suppliers certified to ISO 14001 as one of the company's sustainable development management indicators. By the end of 1999,56 companies had passed ISO 14001 certification. BP-Amoco put forward the goal of obtaining ISO 14001 certification for 131 workplaces by the end of 2001 and all ISO 14001 certification by the end of 2002. Statoil also took ISO 14000 certification in 2005 as a control indicator for the company to create a world-class HSE management system. Passing the ISO 14001 environmental management system certification can promote the continuous improvement of the existing HSE management system. For ISO 14000 certification on the basis of the HSE management system, neither the introduction of an independent environmental management system nor the separation of the environmental management part from the HSE management system will have good results, because the former approach will disrupt the existing The HSE management system also results in the dispersion of resources and the overlapping of some management elements. The latter approach will disrupt the current system and is not conducive to integrated management. To this end, the usual practice is to review the current HSF management system, make further supplements, and introduce a quantitative continuous improvement goal for HSE management, so that it meets the requirements of the ISO 14001 management system, thereby integrating the ISO 14001 management system. Integrate into the existing HSE management system. When using the HSE management system as a carrier to carry out the company's external environmental certification, the "risk" described in the HSE management system must be distinguished from the "major environmental factors" under the ISO 14001 category. HSE management systems focus on reducing risk as low as practicable and as low as practicable (ALARP), whereas ISO 14001 focuses on significant environmental aspects; HSE management systems use the concept of hazard, which is derived from hazards and can lead to top events occur; high risk must carry out correlation diagram analysis to ensure that it is reduced to ALARP, while medium risk or small risk does not need to carry out correlation diagram analysis, only need to establish corresponding procedures or provide necessary capabilities and guarantees; ISO 14001 The Department of Environmental Management requires that all environmental factors be systematically evaluated to determine whether the environmental factors are significant.

Preventive measures and action plans should be established for the identified major environmental factors, and a clear environmental management plan should be specified. To establish an ISO 14001 certified HSE management system, the following aspects should be done: Improve environmental awareness. In order to meet the requirements of the ISO 14001 environmental management system, awareness of environmental issues must be raised within the company. To achieve this, four types of training sessions should be established, targeting department managers, team leaders, supervisors and operators. The establishment of ISO 14001 environmental management system reserve specialized personnel database. In order to successfully implement the ISO 14001 certified management system within the company, it is necessary to continue to deepen the understanding of the environmental management system standard itself within the company.

Therefore, we must first select a group of key personnel to participate in the ISO 14001 environmental management system auditor training course for training, and then hold various types of training courses internally through them. These trainings ensure a full company-wide understanding of ISO 14001 and at the same time that there are enough people to support and participate in internal audits within a certain period of time. Screening for environmental factors. To obtain ISO 14001 certification, the control of significant environmental factors is a must. The screening of environmental factors is generally provided by external consultants with technical guidance, and the identification of environmental factors involves the participation of all employees of the company. Identified major environmental factors should develop feasible preventive measures and action plans, and be subject to internal audit. Select the appropriate review body. Selecting a suitable auditing body to carry out the company's management system certification is the key to making the entire auditing process run smoothly. Auditors should be familiar with the production process of the company and should be geographically close to the company, which can not only ensure the quality of the audit, but also speed up the audit. Keep auditors as consistent as possible from initial to final review, without mandatory requirements. The ISO 14001 certified HSE management system can bring the following benefits: it maintains the consistency with the requirements of the HSE management system, which is a powerful promotion for the HSE management level; it promotes the improvement of employees' environmental awareness and ability, and can quickly generate Effectiveness; adopting an integrated management method for major risks and environmental factors, which is convenient for grass-roots operations; combining ISO 14001 elements with HSE management system elements requires only a small additional cost.

C Establishment of QHSE integrated management system

At present, there are various management system standards in the world, such as the ISO 9000 quality management system of the International Organization for Standardization, the ISO 14000 environmental management system, in addition to the HSFE management system and the occupational health and safety management system formulated by some countries and regions. The implementation of the system has indeed played a positive role in the standardized management of health, safety, and environment, but the existence of multiple systems has also caused this problem. Because these systems are parallel to each other and perform their own duties, the documents, implementation methods and audit processes of different systems are different from each other, which not only brings many difficulties to the operation of the system, but also reduces the function of the system. At the same time, if several independent systems are held, many documents and resources will be reused, which will make the system difficult to control and implement and reduce the efficiency of the enterprise. These problems have aroused the attention of the authors of world standardization. In practice, enterprises have also put forward new countermeasures, such as integrated environment, quality, health, and safety management system. The integrated QHSF management system mainly integrates some common points from the coexisting management systems, such as policy formulation, regulatory compliance, training, document control, supervision, non-conformity control, audit and management review, etc. ; while for some special elements. It is classified and controlled, such as processing, storage and transportation, packaging, services, and inspection of products and services in the quality management system; risk reduction or emergency management control in the health, safety and environmental management system.

Through the implementation of the integrated QHSE management system, the environment, quality, health and safety management system can be simplified, but there are also certain difficulties. The benefits include better integration of QHSE management by operational departments; centralized allocation of resources for environmental, quality, health and safety management; better relationships with government environmental control agencies and other interested parties; improved emergency response contingency planning; higher overall QHSE awareness among workers; open lines of communication with employees; fostering a culture of accepting external QHSE audits; enhancing QHSF performance and corporate functioning. The existing problems are that the management system has changed a lot, and the documents have increased, which needs to be updated and revised continuously; the disposal and elimination of non-conformity requires higher initial investment; the change or increase of the process can cause the change of the system; accidents or some serious operational defects can have an impact on the internal reliability of the system.

7 Risk Factors and Identification Methods

7.1 Definition of hazardous and harmful factors

Risk factors refer to factors that cause casualties to people or sudden damage to objects. Harmful factors refer to factors that can affect human health, cause disease, or cause chronic damage to things. Under normal circumstances, these two are not distinguished and collectively referred to as hazard factors.

Risk or harmful factors mainly refer to equipment, facilities and places with objectively existing dangerous and harmful substances or energy exceeding a certain limit. Although the manifestations of dangerous and harmful factors are different, in essence, the reason why risk or harmful consequences can be caused can be attributed to the combined effects of the existence of risk or harmful substances, energy and risk or harmful substances, and the loss of control of energy. And lead to the leakage, emission and accidental release of energy of dangerous and harmful substances. So the root cause of accidents is the presence of hazards, hazardous substances, energy and out of control.

The incontrollable of hazard substances and energy is mainly reflected in unsafe human action, unsafe status of articles, and management defect. According to GB 6441—86 "Enterprise Employee Casualty Accident Classification", unsafe human action is divided into 13 categories: Operation error; Causing safety device failure; Using unsafe equipment; Hand instead of tool operation; Object storage improperly; Venturing into dangerous places; Climbing or sitting in unsafe positions; Working or staying under lifting objects; Repairing; Checking; Adjusting; Welding; Cleaning.

The deficiencies in management are as follows: Failure to provide safety education and technical training to employees as required, or to operate without passing the job type examination; Lack of safety operation procedures or imperfections; Lack of or defective safety measures, safety signals, safety signs, safety utensils and personal protective equipment; Lack of inspection or wrong guidance for on-site work.

Unsafe status of articles is divided into 4 categories: Lack of or defective equipment, facilities, tools, accessories, etc.; Defective personal protective equipment; Bad environment of production (construction) site; Management deficiencies include safety management system, labor organization, daily safety management and other aspects.

7.2 Classification of hazardous and harmful factors

On GB 6441—86 "Classification of Casualty Accidents of Enterprise Employees", the risk factors are divided into 16 categories by comprehensively considering the causative factors, the leading causes of accidents, harmful substances, and injury methods. They are: object strike; vehicle damage; mechanical damage; lifting injury; electric shock; drowning; burns; fire; suffocation; other injury.

The direct cause of the accident, according to the "Classification and Code for the Hazardous and Harmful Factors in Production Process" (GB/T 13861—92), is divided into 6 categories: Physical hazards and harmful factors; Chemical hazards and harmful factors; Biological hazards dangerous and harmful factors; Psychological and physical hazards and harmful factors; Behavioral hazards and harmful factors; Other dangerous and harmful factors.

According to "Classification of Casualty Accidents for Enterprise Employees" (GB 6441—86), the accident is divided into 20 categories, including object strike, vehicle damage, mechanical damage, lifting damage, electric shock, drowning, burns, fire, fall from height, collapse, roof gang, water seepage, gun explosion, gunpowder explosion, gas explosion, boiler explosion, container explosion, other explosions, poisoning and suffocation, other injuries.

According to the classification of occupational health, with reference to the "Regulations on the Scope of Occupational Diseases and Measures for the Treatment of Occupational Disease Patients" issued by the Ministry of Health, the former Ministry of Labor, and the Federation of Trade Unions, it is divided into risk and harmful factors. They are divided into productive dust, poison, noise and vibration, high temperature and low temperature, radiation (ionizing radiation, non-ionizing radiation) and other harmful factors. The principles that should be followed in the identification of dangerous and harmful factors: scientific, systematic, comprehensive, and predictive. For risk and harmful factors, the triggering events should also be analyzed, that is, the conditions or possible accident modes for the occurrence of risk and harmful factors should be analyzed. Problems that should be paid attention to in the identification of risk and harmful factors should be avoided according to the factory site, plane layout, buildings, materials, production technology and equipment, auxiliary production facilities (including public works), operating environment, etc. Carry out key analysis on direct causes and induced causes, so as to provide a basis for determining evaluation objectives, evaluation priorities, dividing evaluation units, selecting evaluation methods and taking control measures. More importantly, it is necessary to analyze the risk and harmful factors that may cause serious consequences of equipment and device damage and operation errors.

Identification of risk and harmful factors: identification of risk and harmful factors of equipment; identification of risk and harmful factors of working environment; identification of risk and harmful factors related to manual operation; identification of risk and harmful factors during transportation; construction and demolition process identification of risk and harmful factors; identification of risk and harmful factors in the production process. Identification of risk and harmful factors of process equipment and installations identification of risk and harmful factors of chemical and mechanical processing equipment; identification of risk and harmful factors of electrical equipment; identification of risk and harmful factors of special machinery; identification of risk and harmful factors of boilers and pressurevessels; identification of risk and harmful factors of climbing devices.

Hazardous chemicals refer to chemicals that are flammable, explosive, toxic, and harmful, and will cause harm or damage to people, facilities, and the environment. Major hazard source of hazardous chemicals refers to the long-term or temporary production, processing, use or storage of hazardous chemicals, and the quantity of hazardous chemicals equals or exceeds the critical number of units. A unit refers to one (set) of production devices, facilities or sites, or several (sets) of production devices, facilities or sites that belong to the same production and business unit and whose edge distance is less than 500m. Identification basis of major hazard sources of hazardous chemicals: hazardous characteristics of hazardous chemicals, quantity of hazardous chemicals. Identification indicators of major hazard sources: a single hazardous chemical, if the quantity of hazardous chemicals in the unit is equal to or exceeds the critical amount, it is a major hazard source.

References

[1] Wu Qiong. Safety Management [M]. Beijing: Coal Industry Press, 2016.

[2] Jing Guoxun. Principles of Safety [M]. Beijing: National Defense Industry Press, 2014.

[3] Zhang Jinglin. Safety System Engineering [M]. Beijing: Coal Industry Press, 2013.

[4] Lin Baiquan. Principles of Safety [M]. Beijing: Coal Industry Press, 2013.

[5] Zhang Chuanyi, Li Quan. Research on Safety Culture Construction [M]. Xuzhou: China University of Mining and Technology Press, 2012.

[6] Yang Jihua. Brief Introduction to Safety Management [M]. Guangzhou: Guangdong Economic Publishing House, 2012.

[7] Wang Guisheng. Case Analysis of Work Safety Accidents [M]. Beijing: China Architecture & Building Press, 2012.

[8] Miao Jinming, Han Rubing. Modern Enterprise Safety Management Methods and Practice [M]. Beijing: Tsinghua University Press, 2011.

[9] Shi Shaohua. General Introduction to the Rule of Law in Work Safety [M]. Beijing: Coal Industry Press, 2011.

[10] Jin Longzhe, Yang Jixing. Principles of Safety [M]. Beijing: Metallurgical Industry Press, 2010.

[11] Li Feilong. Construction and Implementation of Safety Culture [M]. Beijing: China Labor and Social Security Publishing House, 2010.

[12] Tian Shuicheng, Jing Guoxun. Safety Management [M]. Beijing: China Machine Press, 2009.

[13] Jiang Juncheng. Accident Investigation and Analysis Technology [M]. Beijing: Chemical Industry Press, 2009.

[14] Luo Yun, Cheng Wuyi. Modern Safety Management [M]. Beijing: Chemical Industry Press, 2009.

[15] Yuan Changming, Xie Zhengwen, et al. Safety Management [M]. Beijing: China Metrology Press, 2009.

[16] Luo Yun. Introduction to Safety Science [M]. Beijing: China Quality Inspection press, 2013.

[17] Wu Chao, Wang Ting, et al. Safety Statistics [M]. Beijing: China Machine Press, 2014.

[18] Fu Gui. Safety Management-Behavior Control Method of Accident Prevention [M]. Beijing: Science Press, 2013.

[19] Luo Yun. Risk Analysis and Safety Evaluation (Third Edition) [M]. Beijing: Chemical Industry Press, 2016.

[20] Zhao Huanjuan, Qian Xinming, et al. One-piece Coal Mine Mobile Refuge Chamber with Safety Structure and Less Sealing Risk Based on FEA [J]. Journal of Beijing Institute of Technology, 2014(2): 152-157.

[21] Zhao Huanjuan, Yao Huimei, et al. Experimental Study on the Influence of Initial Pressure on the Structure of Divergebt Cylindrical Detonation Waves [J]. Safety, Health and Environment, 2021, 22(3): 22-27.

[22] Gao Yukun, Zhao Huanjuan, et al. Experimental Teaching Development Research Based on Detonation Pipeline System [J]. China Modern Educational Equipment, 2022, 3: 57-59.

[23] Yan Yiran, Zhao Huanjuan, et al. Experimental Research on Spinning Detonation Structure [J]. Journal of Propulsion Technology, 2021, 42: 593-600.

[24] Niu Shuzhen, Zhao Huanjuan, et al. Mechanism of Pre-mixed CH_4-O_2 Mixture Detonation Propagation in Annular Channel [J]. Journal of Propulsion Technology, 2020, 41(7): 1536-1543.

[25] Yan Yiran, Zhang Yinghua, et al. Effects of Pre-mixed Gas Instability on Sprial Detonation Inner Structure [J]. Journal of Propulsion Technology, 2018, 39(11): 2508-2514.

[26] Niu Shuzhen, Zhao Huanjuan, et al. Experimental Observations of Sprial Detonation Inner Structure [C]. Proceedings of the Chinese Congress of Theoretical and Applied Mechanics, 2019: 2359-2367.

[27] Zhao Huanjuan, Yan Yiran, et al. Experimental Observations of Sprial Detonation Inner Structure in $2H_2$+O_2+3Ar Premixed Gas [J]. Journal of Engineering Thermophysics, 2018, 39(4): 922-929.

[28] Zhao Huanjuan, Bo Yulan, et al. High-frequency Sprial Detonation End-on Records of Pre-mixed C_2H_2+

2. $5O_2$ with Different Argon Dilution [J]. Explosion and Shock Waves, 2018, 38(5): 1121-1129.

[29] Gao Yukun, Wang Shuyi, et al. Experimental Piping Systems for Recording the Internal Structure of Detonation [J]. Research and Exploration in Laboratory, 2017, 36(11): 48-50.

[30] Zhao Huanjuan, Yan Yiran, et al. Experimental Studying on Mach Reflection of Detonation Wave of Premixed Mixtures [J]. Journal of Propulsion Technology, 2017, 38(11): 2572-2579.

[31] Zhao Huanjuan, Yan Yiran, et al. Experimental Observations of Sprial Detonation Inner Structure [C]. Proceedings of the Chinese Mechanics Congress 2017 and Celebrating the 60th Anniversary of the Founding of the Chinese Mechanics Society, 2017: 1059-1065.

[32] Zhao Huanjuan, Lee J H S, et al. Effect of Argon Dilution on Detonation Instability of C_2H_2-O_2 Mixture and Its Quantitative Analysis [J]. Explosion and Shock Waves, 2017, 37(4): 577-584.

[33] Zhao Huanjuan, Lee J H S, Zhang Yinghua, et al. Precise Mechanism of Triple Point Passage Removing Soot on Soot-coated Surface [J]. Journal of Engineering Science, 2017, 37(4): 577-584.

[34] Zhao Huanjuan, Lee J H S, et al. Effects of Boundary Conditions on Premixed $CH_4 + 2O_2$ Detonation Characteristics [J]. Explosion and Shock Waves, 2017, 37(2): 201-207.

[35] Zhao Huanjuan, Lee J H S, et al. Quantitative Irregularity Analysis for Spinning Detonation of Premixed CH_4+2O_2 [J]. Journal of Engineering Science, 2016, 38(11): 1522-1531.

[36] Liu Jie, Zhao Huanjuan, et al. Experimental on Non-self similar of Mach Reflection of Gaseous Detonation Wave [J]. Journal of Aerospace Power, 2016, 31(3): 588-597.

[37] Zhao Huanjuan. Study on Characteristics of Gas Explosion and Optimal Design for Refuge Chamber Under Coal Mine [D]. Beijing: Beijing Institute of Technology, 2015.

[38] Zhao Huanjuan, Yan Yiran, et al. Quantitative Effect of Ar Dilution on the Dynamic Detonation Parameters in Acetylene-Oxygen Mixtures [J]. Advances in Mechanical Engineering, 2016, 10(5): 1-11.

[39] Zhao Huanjuan, Yan Yiran, et al. Propagating Characteristic of Premixed Methane-Oxygen Deflagration in the Coal Mine Lane Including a Refuge Chamber [J]. Journal of Beijing Institute of Technology, 2018, 27(1): 109-117.

[40] Zhao Huanjuan, Lee J H S, et al. Quantitative Comparison of Cellular Patterns of Stable and Unstable Mixtures [J]. Shock Waves, 2016, 26(5): 621-633.

[41] Sun Xuxu, Li Quan, et al. Experimental Investigation of Detonation Propagation in Hydrogen-Air Mixtures in a Tube Filled with Bundles [J]. Experimental Thermal and Fluid Science, 2019, 102: 316-324.

[42] Zhang Bo, Pang Lei, et al. Measurement and Prediction of Detonation Cell Size in Binary Fuel Blends of Methane/Hydrogen Mixtures [J]. Fuel, 2016, 172: 196-199.

[43] Zhang Bo, Liu Hong. The Effects of Large Scale Perturbation-Generating Obstacles on the Propagation of Detonation Filled with Methane-Oxygen Mixture [J]. Combustion and Flame, 2017, 182: 279-287.

[44] Zhang Bo, Wang Cheng, et al. Velocity Fluctuation Analysis Near Detonation Propagation Limits for Stoichiometric Methane-Hydrogen-Oxygen Mixture [J]. International Journal of Hydrogen Energy, 2016, 41(39): 17750-17759.

[45] Zhang Bo. The Influence of Wall Roughness on Detonation Limits in Hydrogen-Oxygen Mixture [J]. Combustion and Flame, 2016, 169: 333-339.

[46] Zhang Bo, Pang Lei, et al. Detonation Limits in Binary Fuel Blends of Methane/Hydrogen Mixtures [J]. Fuel, 2016, 168: 27-33.

[47] Zhang Bo, Liu Hong, et al. An Experimental Study on the Detonability of Gaseous Hydrocarbon Fuel-Oxygen Mixtures in Narrow Channels [J]. Aerospace Science and Technology, 2017, 69: 193-200.

[48] Lee J H S. The Detonation Phenomenon [M]. London: Cambridge University Press, 2008.

[49] Wang Yuhui, Le Jialing. Rotating Detonation Engines with Two Fuel Orifice Schemes [J]. Acta Astronautica,

2019,161:262-275.

[50] Teng Honghui,Jiang Zonglin. Research Progress of Oblique Detonation Multi-Wave Structure and Its Stability [J]. Advances in Mechanics,2020,50:50-92.

[51] Liu Wei,Weng Chunsheng,et al. Initial Flow Field Characteristics of Gas Jet in Underwater Single Detonation of Pulse Detonation Engine [J]. Acta Armamentarii,2020,41(S1):104-109.

[52] Li Jian,Ren Huilan,et al. Length Scale Effect on Mach Reflection of Cellular Detonations [J]. Combustion and Flame,2018,189:378-392.

[53] Liu Jie. Study on Mach Reflection Transition Mechanism of Gas Detonation Wave [D]. Nanjing: Nanjing university of Science and Technology,2016.

[54] Qiangqiang Hao, Zhenmin Luo, et al. The Flammability Limits and Explosion Behaviours of Hydrogen-Enriched Methane-Air Mixtures [J]. Experimental Thermal and Fluid Science,2021,126:1-11.

[55] Shih H Y,Liu C R. A Computational Study on the Combustion of Hydrogen/Methane Blended Fuels for a Micro Gas Turbines [J]. Int J Hydrogen Energy,2014,39(27):15103-15115.

[56] Emadi M,Karkow D,et al. Flame Structure Changes Resulting from Hydrogen-Enrichment and Pressurization for Low-Swirl Premixed Methane-Air Flames [J]. Int J Hydrogen Energy,2012,37(13):10397-10404.

[57] Bouras F,El M,et al. Control of Methane Flame Properties by Hydrogen Fuel Addition:Application to Power Plant Combustion Chamber [J]. International Journal of Hydrogen Energy,2017,42(13):8932-8939.

[58] Parag Rajpara,Rupesh Shah,et al. Effect of Hydrogen Addition on Combustion and Emission Characteristics of Methane Fuelled Upward Swirl Can Combustor [J]. International Journal of Hydrogen Energy, 2018, 43(36):17505-17519.

[59] Burbano H J, Amell A A, et al. Effects of Hydrogen Addition to Methane on the Flame Structure and CO Emissions in Atmospheric Burners [J]. Int J Hydrogen Energy,2008,33(13):3410-3415.

[60] Koç Y,Yağlı H,et al. Analysing the Performance,Fuel Cost an Demission Parameters of the 50MW Simple and Recuperative Gas Turbine Cycles using Natural Gas and Hydrogen as Fuel [J]. Int J Hydrogen Energy, 2020,45(41):22138-22147.

[61] Cellek M S,Pınarbaşı A. Investigations on Performance an Demission Characteristics of an Industrial Low Swirl Burner While Burning Natural Gas,Methane,Hydrogen-Enriched Natural Gas and Hydrogen as Fuels [J]. Int J Hydrogen Energy,2018,43(2):1194-1207.

[62] Taamallah S,Vogiatzaki K,et al. Fuel flexibility,Stability and Emissions in Premixed Hydrogen-Rich Gas Turbine Combustion:Technology,Fundamentals,and Numerical Simulations [J]. Applied Energy,2015,154: 1020-1047.

[63] Taamallah S,LaBry Z A,et al. Thermo-Acoustic Instabilities in Lean Premixed Swirl-Stabilized Combustion and Their Link to Acoustically Coupled and Decoupled Flame Macrostructures [J]. Proceedings of the Combustion Institute,2015,35(3):3273-3282.

[64] Zhang Jianan,Ratner Albert. Experimental Study of the Effects of Hydrogen Addition on the Thermoacoustic Instability in a Variable-Length Combustor [J]. International Journal of Hydrogen Energy,2021,46(29): 16086-16100.

[65] Gorkem Oztarlik,Laurent Selle,et al. Suppression of Instabilities of Swirled Premixed Flames with Minimal Secondary Hydrogen Injection [J]. Combustion and Flame,2020,214:266-276.

[66] Kim H S,Arghode V K,et al. Hydrogen Addition Effects in a Confined Swirl-Stabilized Methane-Air Flame [J]. International Journal of Hydrogen Energy,2009,34(2):1054-1062.

[67] Laera D,Agostinelli P W,et al. Stabilization Mechanisms of CH_4 Premixed Swirled Flame Enriched with a Non-Premixed Hydrogen Injection [J]. Proceedings of the Combustion Institute,2021,38(4):6355-6363.

[68] Gorkem Oztarlik, Laurent Selle, et al. Suppression of Instabilities of Swirled Premixed Flames with Minimal Secondary Hydrogen Injection [J]. Combustion and Flame, 2020, 214: 266-276.

[69] Chaudhuri S, Kostka S, et al. Blowoff Dynamics of Bluff Body Stabilized Turbulent Premixed Flames [J]. Combust Flame, 2010, 157(4): 790-802.

[70] Hayakawa A, Arakawa Y, et al. Experimental Investigation of Stabilization and Emission Characteristics of Ammonia/Air Premixed Flames in a Swirl Combustor [J]. Int J Hydrogen Energy, 2017, 42 (19): 14010-14018.

[71] Guo S, Wang J, et al. Effect of Hydrogen Enrichment on Swirl/Bluff-Body Lean Premixed Flame Stabilization [J]. Int J Hydrogen Energy, 2020, 45(18): 10906-10919.

[71] Mao Runze, Wang Jinhua, et al. Effect of High Hydrogen Enrichment on the Outer-Shear-Layer Flame of Confined Lean Premixed CH_4/H_2/Air Swirl Flames [J]. International Journal of Hydrogen Energy, 2021, 46(34): 17969-17981.

[73] Huang Zuohua, Liu Bing, et al. Combustion Characteristics and Heat Release Analysis of a Spark-Ignited Engine Fueled with Natural Gas-Hydrogen Blends [J]. Energy & Fuels, 2007, 21(5): 2594-2599.

[74] Wen Xiaoping, Wang Mengming, et al. Suppression Effects of Ultrafine Water Mist on Hydrogen/Methane Mixture Explosion in an Obstructed Chamber [J]. International Journal of Hydrogen Energy, 2019, 44(60): 32332-32342.

[75] Gao Y, Ng H D, et al. Minimum Tube Diameters for Steady Propagation of Gaseous Detonations [J]. Shock Waves, 2014, 24(4): 447-454.

[76] Wojciech Rudy, Mateusz Zbikowski, et al. Detonations in Hydrogen-Methane-Air Mixtures in semi Confined Flat Channels [J]. Energy, 2016, 116: 1479-1483.

[77] Hoorelbeke P, Mukharror D. Mitigation of Vapour Cloud Explosions Effects with Inhibitors [C]. Asia Pacific Health, Safety Security and Environment Conference, Soc. Petroleum Eng, 2009.

[78] Liang Wenkai, Liu Zirui, et al. Explosion Limits of $H_2/CH_4/O_2$ Mixtures: Analyticity and Dominant Kinetics [J]. Proceedings of the Combustion Institute, 2019, 37(1): 493-500.

[79] Hashemi Hamid, Christensen J M, et al. High-Pressure Oxidation of Methane [J]. Combustion and Flame, 2016, 172: 349-364.

[80] Zhang Y, Huang Z, et al. Experimental and Modeling Study on Ignition Delays of Lean Mixtures of Methane, Hydrogen, Oxygen, and Argon at Elevated Pressures [J]. Combustion and Flame, 2012, 159(3): 918-931.

[81] Sun Zuoyu. Experimental Studies on the Explosion Indices in Turbulent Stoichiometric H_2/CH_4/Air Mixtures [J]. International Journal of Hydrogen Energy, 2019, 44(1): 469-476.

[82] Gao Y, Zhang B, et al. An Experimental Investigation of Detonation Limits in Hydrogen-Oxygen-Argon Mixtures [J]. International Journal of Hydrogen Energy, 2016, 41(14): 6076-6083.

[83] Zhang Bo, Liu Hong, et al. Effect of Acoustically Absorbing Wall Tubes on the Near-Limit Detonation Propagation Behaviors in a Methane-Oxygen Mixture [J]. Fuel, 2019, 236: 975-983.

[84] Wu Yuwen, Zheng Quan, et al. An Experimental Study on the Detonation Transmission Behaviours in Acetylene-Oxygen-Argon Mixtures [J]. Energy, 2018, 143: 554-561.

[85] Wang Luqing, Ma Honghao, et al. Effects of Bluff Bodies on the Propagation Behaviors of Gaseous Detonation [J]. Combustion and Flame, 2019, 201: 118-128.

[86] Rainsford Georgina, Aulakh D J S, et al. Visualization of Detonation Propagation in a Round Tube Equipped with Repeating Orifice Plates [J]. Combustion and Flame, 2018, 198: 205-221.

[87] Mitchell Cross, Gaby Ciccarelli. DDT and Detonation Propagation Limits in an Obstacle Filled Tube [J]. Journal of Loss Prevention in the Process Industries, 2015, 36: 380-386.

[88] Starr Amanda, Lee J H S, et al. Detonation Limits in Rough Walled Tubes [J]. Proceedings of the Combustion Institute, 2015, 35(2): 1989-1996.

[89] Jesuthasan. Near-Limit Propagation of Detonations [D]. Canada: McGill University, 2011.

[90] Lee J H S, Jesuthasan A, et al. Near Limit Behavior of the Detonation Velocity [J]. Proceedings of the Combustion Institute, 2013, 34(2): 1957-1963.

[91] Goodwin G B, Houim R W, et al. Effect of Decreasing Blockage Ratio on DDT in Small Channels with Obstacles [J]. Combustion and Flame, 2016, 173: 16-26.

[92] Mitchell Cross, Gaby Ciccarelli. DDT and Detonation Propagation Limits in an Obstacle Filled Tube [J]. Journal of Loss Prevention in the Process Industries, 2015, 36: 380-386.

[93] Kiumars Mazaheri, Yasser Mahmoudi, et al. Experimental and Numerical Investigation of Propagation Mechanism of Gaseous Detonations in Channels with Porous Walls [J]. Combustion and Flame, 2015, 162(6): 2638-2659.

[94] Teng Honghui, Ng Hoi Dick, et al. Evolution of Cellular Structures on Oblique Detonation Surfaces [J]. Combustion and Flame, 2014, 162(2): 470-477.

[95] Monette M. Detonation Propagation in Rough Tubes [D]. Canada: McGill University, 2014.

[96] Zhou Shangyong, Gao Jiancun, et al. Effects of Mesh Aluminium Alloy and Aluminium Velvet on the Explosion of H_2/Air, CH_4/Air and C_2H_2/Air Mixtures [J]. International Journal of Hydrogen Energy, 2021, 46(27): 14871-14880.

[97] Li Quan, Wei Zhen, et al. The Behaviors of Supersonic Combustion Wave Through a Perforated Plate in a Stoichiometric Mixtures of $H_2/CH_4/O_2$ and H_2/O_2 [J]. Fuel, 2022, 317: 1-9.

[98] Wang C, Huang F L, et al. Effect of Concentration and Obstacles on Flame Velocity and Overpressure of Methane-Air Mixture [J]. Journal of Loss Prevention in the Process Industries, 2016, 43: 302-310.

[99] Song Xianzhao, Zuo Xuechao, et al. The Explosion-Suppression Performance of Mesh Aluminum Alloys and Spherical Nonmetallic Materials on Hydrogen-Air Mixtures [J]. International Journal of Hydrogen Energy, 2020, 45(56): 32686-32701.

[100] Yang Zhongkun, Zhao Kun, et al. Effects of Mesh Aluminium Alloys and Propane Addition on the Explosion-Suppression Characteristics of Hydrogen-Air Mixture [J]. International Journal of Hydrogen Energy, 2021, 46(70): 34998-35013.

[101] Liu Yong, Liu Xiaotian, et al. Numerical Investigation of Hydrogen Detonation Suppression with Inert Particle in Pipelines [J]. International Journal of Hydrogen Energy, 2016, 41(46): 21548-21563.

[102] Li Yanchao, Bi Mingshu, et al. Inerting Effect of Carbon Dioxide on Confined Hydrogen Explosion [J]. International Journal of Hydrogen Energy, 2019, 44(40): 22620-22631.

[103] Chen Yongzheng, Liu Bo, et al. Effects of Nitrogen and Carbon Monoxide on the Detonation of Hydrogen-Air Gaseous Mixtures [J]. Nuclear Engineering and Design, 2019, 343: 1-10.

[104] GB/T 13745—09, Classification and Code of Disciplines[S].

[105] GB/T 13745—92, National Standards Discipline Classification and Code[S].

[106] GB 6441—86, The Classification for Casualty Accidents of Enterprise Staff and Workers[S].

[107] GB 5044—85, Classification for Hazards of Occupational Exposure to Toxicant[S].

[108] GB 12331—90, Classified Standard of Operation with Exposure to Toxic Chemicals in the Workplace[S].

[109] GB 5817—86, Classification of Hazard Levels due to Exposure to Industrial[S].

[110] GBJ 87—85, Code for Design of Noise Control in Industrial Enterprises[S].

[111] GB 4200—97, Classification of Works in Hot Environment[S].

[112] GB/T 14440—92, Classification of Works in Cold Environment[S].

[113] GB/T 13861—92, Classification and Code for the Hazardous and Harmful Factors in Production Process [S].

[114] GB 935—89, Limit Value of Allowable Continuous Heat Exposure Time in Hot Environment[S].